Decolonising Seed Governance in the Andes

Focusing on the Latin American Andes region, this book examines how emerging decolonial narratives, practices, and rules have opened possibilities for structural change in agri-food systems.

The book argues that the contemporary global agri-food crisis is the outcome of a modern and colonial worldview that has reached its limits. Addressing it requires deep transformations related to power, values, and ways of being and knowing. Over the last 35 years, the Andean region has experienced a 'trans-modern turn', planting the seeds for deep transitions towards alternatives like Buen Vivir while challenging established global north paradigms. Using agricultural seeds as the site of analysis, the book reveals how struggles over knowledge, authority, and legitimacy are condensed in everyday practices of seed production, exchange, and regulation. It traces how Andean Seed Guardian Networks engaged the state to gradually reshape seed systems. During the early 21st century, Colombia, Ecuador, and Venezuela became sites of significant institutional experimentation. Resulting reconfigurations of state–society relations led to novel governance arrangements and micro-institutions that challenge extractivist industrialist agriculture while advancing food sovereignty and agrobiodiversity. The book is theoretically plural, drawing on social practice theory, discourse theory, new institutionalism, and sustainability transitions theory to analyse how alternative worldviews are materialised and institutionalised. It contributes to debates on just transitions, epistemic justice, and the decolonisation of global food systems.

This book will be of interest to scholars and practitioners in food studies, sustainability transitions, development studies, political ecology, social innovation, and Latin American studies.

Juan Garzón is an independent consultant, researcher and university lecturer working at the intersection of sustainable transitions and social innovation. He holds a PhD in Sustainable Futures from the University of Technology Sydney, Australia. Juan is currently a lecturer at the Department of Business Administration at the Pontificia Universidad Javeriana, Colombia.

Earthscan Food and Agriculture

Multiple Capitals Approach for Upland Agricultural Systems
Lois Mansfield

Agricultural Land Use
Structural Transformations, Environment Challenges, Planning and Policy
Edited by Jerzy Bański

Sustainable Production of Root and Tuber Crops
Edited by Ravinder Kumar, Milan Kumar Lal, Rahul Kumar Tiwari and Brajesh Singh

Agrifood Regimes in Greece
Sociotechnical Dynamics and Sustainable Transitions
Stathis Arapostathis, Sotiris Alexakis, Vasiliki Karantzavelou and Yannis Fotopoulos

Nourishing Scotland
Pete Ritchie

Community Supported Agriculture in Central and Eastern Europe
Transitional Economies in Post-Communist Europe
Magdalena Raftowicz & Mirosław Struś

Decolonising Seed Governance in the Andes
Lessons for Just and Sustainable Food Futures
Juan Garzón

Nature-Based Solutions for European Vineyards
Solutions for Transformative Change
Edited by Udaya Sekhar Nagothu

American Farmland Preservation
Policies, Practices and the Politics of Place
Andrew Waters

For more information about this series, please visit: www.routledge.com/Earthscan-Food-and-Agriculture/book-series/ECEFA

Decolonising Seed Governance in the Andes

Lessons for Just and Sustainable Food Futures

Juan Garzón

Routledge
Taylor & Francis Group
LONDON AND NEW YORK

First published 2026
by Routledge
4 Park Square, Milton Park, Abingdon, Oxon OX14 4RN

and by Routledge
605 Third Avenue, New York, NY 10158

Routledge is an imprint of the Taylor & Francis Group, an informa business

British Library Cataloguing-in-Publication Data
A catalogue record for this book is available from the British Library

Library of Congress Cataloging-in-Publication Data
A catalog record has been requested for this book

ISBN: 978-1-041-20111-3 (hbk)
ISBN: 978-1-041-20112-0 (pbk)
ISBN: 978-1-003-71509-2 (ebk)

DOI: 10.4324/9781003715092

Typeset in Times New Roman
by Taylor & Francis Books

Contents

Illustrations

Figures

Tables

Acknowledgements

This book presents PhD research supported by an Australian Government Research Training Program scholarship. I am forever indebted to the Australian people.

I would like to express my deepest gratitude to Prof. Chris Riedy and Prof. Cameron Tonkinwise for their guidance and support throughout the journey. Many thanks to friends and colleagues in the transitions and social change group, the Institute for Sustainable Futures, DAB, and TD School at UTS for their support.

This endeavour would not have been possible without the generosity of research participants. I am grateful for their contributions. They have my admiration for their efforts to bring forth more just and sustainable futures.

Words cannot express my gratitude to my family who have offered their love and support through the toughest of times. This work is a product of their efforts.

I would like to extend my sincere thanks to the examiners throughout the PhD journey, to all who contributed insightful comments and suggestions, and the editorial team at Routledge for their support.

Acronyms

ACOSEMILLAS	Colombian Association of Seeds and Biotechnology
AKS	Agricultural knowledge system
ARN	Andean Reciprocal Nurturing
BV	*Buen Vivir*
CAN	Andean Community of Nations
CODECYT	Corporation for Scientific and Technological Development (Venezuela)
CONAIE	Confederation of Indigenous Nationalities of Ecuador
COPISA	Plurinational and Intercultural Conference on Food Sovereignty (Ecuador)
CORPOICA	Colombian Corporation for Agricultural Research
DUS (criteria)	Distinctiveness, uniformity and stability
ESGN	Ecuadorian seed guardian network
FAO-UN	Food and Agriculture Organization of the United Nations
FENOCIN	National Confederation of Peasant, Indigenous, and Black Organisations (Ecuador)
GI	Grassroots innovations
GM	Genetically modified
HYVs	High-yield varieties
ICA	Colombian Institute of Agriculture
ICTs	Information and communication technologies
INIA	National Institute of Agricultural Research (Venezuela)
IPR	Intellectual property rights
ITPGRFA	International Treaty on Plant Genetic Resources for Food and Agriculture
KS	Knowledge systems
LVC	La Via Campesina
MAG	Ecuadorian Ministry of Agriculture

MIA	National Agricultural and Popular Roundtable for Dialogue and Agreement (Colombia)
MLP	Multi-level perspective
ONDB	National Office of Biodiversity (Venezuela)
ONIC	National Indigenous Organization of Colombia
PBR	Plant breeder rights
PFA	Peasant and family agriculture
PFCA	Peasant, family, and community agriculture
PVR	Plant variety rights
RA	Regenerative agriculture
SCPM	Superintendency of Control over Market Power (Ecuador)
SGNs	Seed guardian networks
SPT	Social practice theory
STS	Socio-technical systems
ST	Sustainability transitions
TCZ	Translational contact zones
TRIPS	Trade-Related Intellectual Property Rights agreement
UNDRIP	Declaration on the Rights of Indigenous Peoples (2007)
UNDROP	Declaration on the Rights of Peasants and Others Working in Rural Areas (2018)
UNORCAC	Union of Peasant and Indigenous Organisations of Cotacachi (Ecuador)
UOSISP	Union of Intercultural Social Organisations of Pichincha (Ecuador)
UPOV	International Union for the Protection of New Varieties of Plants

Glossary

Abyssal line	A conceptual line produced by Modernity/Coloniality dividing the Modern and non-Modern. Modernity/Coloniality invents colonial difference between a superior subspecies and all other (non)humans. The universalisation of this totality produces the phenomenon of abyssal thinking, the impossibility of co-presence of both sides of the abyssal line.
AfroIndopeasant	Refers to Afro-Latino, Indigenous, and peasant communities of Latin America. This term does not intend to subsume these groups under a single monolithic identity. Rather, it emphasises their shared non-Modern narratives and practices.
Agroecology	The application of ecological principles to localised food systems while prioritising knowledge and agency of Indigenous and peasant actors. It is simultaneously science, practice, and social movement, as it is also politicised to transform food systems.
Anthropocene	An ecological event produced by the aggregated effects of human activities leading to transgressive and enduring modifications of biophysical earth systems.
Anthropo-not-seen	The process by which alternative life-worlds that do not separate humans from non-humans are forced into that distinction.
Buen Vivir	'Good living' in English; it is a conceptual alternative to Western development focused on living in harmony with oneself, community, and nature. The term has Amazonian Indigenous origins but has been mainstreamed in Latin America and is used as a future vision to organise processes at multiple scales.

Coloniality	The recognition that Modernity and Colonialism are a single entity producing a matrix of power organising the world in core-periphery arrangements. It positioned whiteness, industrialism, capitalism, anthropocentrism, rationalism, and other isms at the core, excluding all those who are non-conforming.
Conuco	A traditional Latin American agroforestry system based on small parcels generally connected to a family home, polycultures, and family agriculture.
Development	The globalisation of Westernisation as a process of social change towards homogenisation by an unilinear understanding of history, with industrialism as its inevitable end.
Degrowth	The reduction of production, consumption, and accumulation to enhance wellness as a response to the environmental and social impacts of economic growth and associated material intensity.
Dialogue of knowledge(s)	The act of finding common terms and engaging in dialogue with plural forms of knowledge, particularly with those emerging from worldviews peripheral to Modernity/Coloniality.
Discourse	Shared ways of looking at the world. Specific ensembles of ideas, concepts, and categorisations giving meaning to physical and social realities. They construct relationships helping to define commonsense and legitimate knowledge.
Discourse coalition	Ensembles of actors attracted to specific storylines and reproducing them through practices.
Eco-socialism	A socio-political approach that links social equality and environmental sustainability by challenging tenets of capitalism.
Epistemic	Related to epistemology.
Epistemology	How we know or understand what our reality is.
Family agriculture	A form of small-scale production where family labour is central, including farm management and operation. Generally passed down through generations.
Food sovereignty	The right of people to produce and consume tasty, healthy, and culturally relevant food produced sustainably. Emphasises the reconfiguration of social relations to empower historically marginalised actors.

Germplasm	Living biological materials carrying genetic information of a plant. Needed to grow them and pass their traits (e.g., appearance) to future generations.
Green revolution	A global design of agricultural development based on extensive monocultures needing improved seeds, sophisticated machinery, and large amounts of agrichemicals.
Global North	Refers to communities, ideas, practices, etc. aligned with the core of Modernity/Coloniality.
Global South	Refers to communities, ideas, practices, etc. pushed to the peripheries of Modernity/Coloniality.
Knowledge(s)	Refers to plural forms of knowledge. The (s) is added to emphasise historically excluded knowledge.
Milpa	An ancestral Meso-American method of cyclical cultivation involving diverse crops and more spiritual modes of relation between farmers and land.
Modern(ity)/Colonial(ity)	The indivisibility of Modernity and Colonialism as a global organising force arranging the world according to levels of whiteness, industrialism, capitalism, anthropocentrism, rationalism, heteronormativity, ableism, and other categories. Used interchangeably with 'coloniality of power' and the 'colonial matrix of power'.
Modern	Refers to communities, ideas, practices, etc. aligned with the core of Modernity/Coloniality (e.g., individualism, Euro-centred white superiority, extractivism).
Multi-level perspective	A process theory modelling socio-technical transitions as resulting from interactions between processes at different levels: micro (niche), meso (regime), and macro (landscape). For a transition to happen, niches at the periphery of the regime articulate with landscape pressures to destabilise and transform the regime.
Non-Modern	Refers to communities, ideas, practices, etc. pushed to the peripheries of Modernity/Coloniality but very much still alive (e.g., humans as part of nature, Indigenous knowledge).
Ontology	What is our reality and how existing things are related to each other.
Other(ed)	Refers to communities, ideas, practices, etc. pushed to the peripheries of Modernity/Coloniality, particularly as they are understood as deviant despite not being harmful.

Peasant	*Campesino* in Spanish. Individuals or communities engaging in small-scale agricultural production for subsistence and/or markets, relying on some family and other non-monetised forms of labour, and having some attachment to their land.
Phytosanitary	Refers to all that relates to plant health, particularly as part of transnational trade.
Pluriverse, pluriversal	Rejection of a homogeneous and monolithic world and the possibility of a world where many worlds fit. It goes beyond multiculturalism, emphasising plurality and horizontal intercultural dialogue.
Power	The capacity or incapacity of actors to mobilise institutions and mental, human, artifactual, natural, or monetary resources to achieve a goal.
Seed sovereignty	The right of peoples to use, save, and exchange seeds and participate in seed governance according to their needs.
Social practices	The source and carrier of meaning, language, and normativity. Practices are articulations between meanings (ideas, aspirations, values), competences (shared know-hows), and materials (technologies, infrastructure).
Sustainability transitions	Long-term, non-linear processes of socio-technical change that last between 25 and 50 years, where interactions between societal subsystems influence the dynamics of individual subsystems, leading to irreversible patterns of change. These are multi-actor, multi-phase, multi-level processes guided by targets/visions, where sustainability is understood as a moving target. Also characterised by collaborative design and learning.
Transgenic seeds	Seeds that have been genetically modified by inserting genes from different species to achieve specific traits.
Translational contact zone	A space where translation can occur across the abyssal line. Translation is understood as a polycentric process prone to equivocations between the Modern and non-Modern as agents, knowledges, and practices interact. It is a recognition of reciprocal incompleteness that allows the expansion of the parameters that define authorship, intelligibility, and purposefulness.

Transmodern(ity)	A minority worldview that aims to become a future culture incorporating the best of the pre-Modern, Modern, and post-Modern to produce a pluriversal utopia through horizontal intercultural dialogue. It is a retelling of human history from those who have been marginalised by Modernity/Coloniality as they construct their own future visions and meta-narratives of social transformation.

Introduction

A consensus is forming globally: the dominant agri-food system is in the midst of a multidimensional crisis. Calls to transform it are increasing in quantity and urgency as it threatens food security on a planetary scale. The impacts of the climate crisis, shifting demographics, and deepening inequalities are placing immense pressure on the systems that feed us. Mere technological improvements, isolated behavioural changes, or superficial policy reforms will be insufficient to meet existing pressures, let alone those to come. The magnitude of the environmental and social crises humanity faces requires something far more radical: deep structural transformations related to power, values, and worldviews.

This book focuses on those deep structures in global agri-food systems, attempting to evidence how they manifest, lock-in, and occasionally crack. It frames the current configuration of the dominant food system not as an inevitability but a product of a Modern/Colonial worldview; a power structure that appears to have exhausted itself, failing to provide viable solutions to the predicaments it has produced. Accordingly, the book aligns with the decolonial dictum that there are no Modern solutions to Modern problems. Humanity must transition towards more radical alternatives.

While there is a growing interest in food system transformation in academic and policy circles, there is limited analysis of how such changes happen on a large scale, particularly outside rich nations. This book aims to partly fill this gap by analysing how transformation occurred in Andean seed systems during the early 21st century. This region and timeframe were selected because they represent a unique historical juncture where alternative worldviews successfully informed new and more inclusive governance configurations.

Andean seed systems are particularly illuminating because they are hybrids between industrial and native seeds, between corporate science and ancestral practice, and between the Modern/Colonial and its peripheries. In the early 21st century Andes, eco-socialism, Indigenous concepts of wellbeing (*Buen Vivir*), direct democracy, and food sovereignty reshaped legal frameworks and system relationships. Diverse worldviews interacted, clashed, and engaged in creative tensions to produce new food futures.

DOI: 10.4324/9781003715092-1

Seeds as a vehicle for enquiry

Agricultural seeds are ideal to explore tensions between worldviews. They are fluid to the point of becoming puzzling. They resist any single, static definition because narratives about them are as varied as they are contradictory, multiple practices revolve around them, and they traverse multiple scales in time and space (from ancestral genes to global supply chains).

What, then, are seeds? Are they national strategic resources to be protected by the state? Or are they collective human heritage and free, living beings with agency? Are they genetic codes to be improved, and patented by science, or are they kin to be loved, nurtured, and respected? Are they standardised commodities subjected to the logic of planned obsolescence, or are they worthy of spiritual reverence as the origin of life itself?

They seem to be all these and more.

This fluidity extends to the concept of agrobiodiversity. Where is it located? Is it found in the seeds themselves, in the germplasm frozen in the vaults of government, academic, or private seed banks? Is it hiding in distant places where human reach is limited? Is it embodied in the calloused hands and memories of Indigenous and peasant farmers, or in the databases and spreadsheets of scientists? It appears to be in all these places simultaneously.

In the smallest of seeds, worldviews collide and futures are contested. Seeds bring structural tensions to the fore; they act as an axis around which governments, industry, and civil society revolve, attempting to tackle historical issues of knowledge and power. These issues ultimately demand dialogue, creative tension, and antagonism between plural knowledge(s) to co-produce sustainable futures.

A shift in the Andes

Over the last 35 years, there has been a considerable shift in seed system governance in the Andean region. Seed systems have evolved from being exclusionary industrial-centric regimes to attempting to integrate Indigenous and traditional seed knowledge(s).

This book argues that this shift can be interpreted as the kick-off phase of a decolonial pathway towards more regenerative food futures. Although this assertion, and indeed the book's title, might sound overly ambitious, the hope is that readers find sufficient evidence within these pages to convince them that alternative futures are not just possible but are already emerging. Human societies have experienced 500 years of Coloniality, and decolonisation might take 500 more. However, there have always been cracks in established systems of power. As Modern crises worsen, collectives expand these cracks and leverage openings to advance alternatives with increasing success.

Drawing from cases of seed system transformation in Colombia, Ecuador, and Venezuela, this book analyses where and how change happened. It attempts to synthesise lessons that might be applicable in other contexts. It draws from diverse theories including sustainability transitions, new institutionalism, social practice theory, and discourse theory. In this sense, the book is intellectually promiscuous. However, it does not attempt to integrate theories; rather, it uses them as distinct lenses to better understand the multifaceted nature of change.

Agri-food systems and large systems change research are massive, dynamic fields that touch upon many variables from a multitude of disciplines. As such, this book makes no claim to be exhaustive. Instead, it humbly narrates the story of seed sovereignty social movements encountering Andean states to produce change. It tells the story of how, by amplifying citizen participation while being influenced by alternative narratives, these actors expanded their understandings of what is valuable, legitimate, and possible in national food systems.

A transition from a degenerative and extractive food system to a regenerative one is urgent, yet the guiding visions and pathways remain contentious. This book shows how such processes have been operationalised in the Andean region, with the hope that this contributes to better understanding and facilitating them globally. It is a contribution to the decolonisation of global food systems, in search for the alternatives we so desperately need to solve our current crises.

Structure of the book

The book is divided into three distinct parts, designed to guide the reader from the theoretical foundations to the empirical realities, and finally to a synthesis of how change occurred.

Part 1: Modernity/Coloniality and Transmodernity

Chapter 1 introduces the concepts of Modernity/Coloniality and the abyssal line as the pre-eminent structures that must be transformed. Coloniality refers to the recognition of Modernity and Colonialism not as separate historical epochs but as a single, intertwined entity; a matrix of power that establishes core-periphery dynamics dominated by anthropocentrism, capitalism, industrialism, and rationalism. It produces an abyssal line dividing the Modern core from its non-Modern peripheries. This dynamic renders the non-Modern invisible, as Modern learning can only occur in colonial terms. The chapter argues that these structures are starkly evident in contemporary seed systems, perpetuating multiple injustices by systematically discriminating against an 'informal' seed system.

Chapter 2 introduces a broad transformative alternative by articulating the concepts of Transmodernity, *Buen Vivir*, and translational contact zones. Transmodernity refers to a worldview emerging from the margins, aiming to become a future culture that assumes the best of the pre-Modern, Modern, and post-Modern to produce a pluriversal[1] utopia through horizontal intercultural dialogue. It is a retelling of human history from the perspective of those who have been marginalised by the colonial matrix of power.

Buen Vivir is a key component of this alternative. It is a Latin American proposal of wellbeing focused on living in harmony with oneself, community, and nature. This concept, with Amazonian Indigenous origins, has been mainstreamed to a point where it now informs transformations at a national scale. Transmodernity and *Buen Vivir* can be enacted through the opening of translational contact zones, spaces where translation occurs between the Modern and non-Modern. In its most basic form, this translation is a recognition of reciprocal incompleteness, allowing for the expansion of the parameters that define authorship, intelligibility, and purposefulness.

Part 2: Andean seed system change

Part 2 defines the contours of seed sovereignty social movements in the region while connecting them to global processes of resistance against the dominant food regime. Establishing these social movements as key actors, it then presents three distinct processes of seed system change in Colombia, Ecuador, and Venezuela.

Chapter 3 provides thick descriptions of the narratives and practices of Latin American seed guardian networks (SGNs). SGNs defend the rights of peoples to produce, store, improve, exchange, and trade seeds according to Indigenous, Afro-descendant, peasant,[2] family,[3] agroecological, and organic agricultural narratives and practices. They can be described as diverse, decentralised, place-based, trans-local coalitions of people and organisations committed to safeguarding care practices of native seeds.

Narratively, seed guardians understand seeds as the starting point of nourishment, but also as free, living entities that embody the origin and regeneration of existence. For them, humans and seeds share a deep relationship of nurturing and care, a connection that mobilises stories of seeds as part of family and community. Guardians understand seeds as an intergenerational commons beyond government and corporate control, sometimes going as far as speaking of seeds as beings with agency and rights of their own.

Seed guardians are widely diverse but engage in two main practices: guardianship and exchange. Seed guardianship relates to the in-situ conservation of seeds that are viable in a specific territory. Guardians may gift or sell seeds to people interested in growing different crops or may exchange them by bartering with other farmers in specialised events. Seed guardians also meet virtually and physically to share knowledge, discuss the current state of seed systems, organise politically, and engage in collective action.

SGNs started as a response to the intensification of neoliberal policies that opened more spaces for transnational corporations while curtailing possibilities for traditional agricultural practices. Initially, their activities were directed towards resisting legislative changes and simply getting their voices heard. Over time, however, SGNs have been able to consolidate, effectively creating protected spaces for innovation and eventually transforming seed system rules. SGNs have connected otherwise isolated outliers to simultaneously resist the dominant system and propose viable alternatives.

The networks have been able to open small cracks in the dominant agri-food regime, despite a pervasive industrialist bias that seeks to exclude them. The decade of 2015–2025 has been full of important developments that show a renewed recognition of traditional knowledge and practices in Andean seed systems. This has been implemented in novel legal frameworks and emerging institutions that set a more levelled playing field between industrial and traditional seed production and distribution.

Chapters 4–6 describe three cases of seed system transformations chronologically.

Chapter 4 describes the emergence of a co-creation process in Colombia, purposefully including Indigenous actors and SGNs to define the national seed system. This comes after more than a decade of tensions due to the signing of international trade partnership agreements and encroaching neoliberal policies. Following the signing of a peace accord between the FARC-EP guerrilla group and the government, timid *Buen Vivir* framings pushed the state to be more inclusive of small-scale and family farming. This influenced reconfigurations of the national seed authority and other agencies, the development of policy guidelines for agroecology, and a national seed plan inclusive of traditional practices.

Chapter 5 describes the process by which Venezuela approved what may be the most comprehensive seed law in the region. Informed by a top-down eco-socialist framing, the law recognises and validates two seed systems (Industrial and AfroIndopeasant), acknowledges place-based community knowledge-holders, and ordains the creation of a scientific–peasant alliance to promote horizontal knowledge exchanges.

Chapter 6 describes the process by which Ecuador developed a seed law that recognises and encourages traditional practices. It emerges from a constitutional process based on principles of *Buen Vivir*, which opens the state to more participatory democracy and directs it towards the attainment of food sovereignty. As a result of this process, new government agencies were formed to promote more horizontal knowledge exchanges between scientists, government officials, and traditional farmers.

Part 3: How change happened

Part 3 re-engages with theory to make sense of Andean seed system change in the early 21st century. It argues that change occurred mainly in three sites: narratives, practices, and rules. It posits that this 35-year process can be interpreted as the kick-off phase of a wider transition in the region. It analyses this process as a socio-technical transition with special interest in issues of power and justice.

Chapter 7 explores narratives, practices, and rules as the main sites where change occurred. It uses the lens of narrative transformation to define the ideational elements that were mobilised throughout the process and explain the formation of discourse coalitions. Practices are seen through the lens of social practice theory, emphasising articulations between meanings, competences, and materials. SGN consolidation is framed as a process of practice governance. It argues that the consolidation of SGNs, along with shifts in state practices, enabled legislative change. These new rules are seen through the lens of new institutionalism to argue that novel micro-institutions emerged and have the potential to further transform the system.

Chapter 8 deploys sustainability transitions theory and its multi-level perspective (MLP) model with a heavy emphasis on power. Seed system transformations are analysed as long-term, non-linear processes of socio-technical change. These processes are described as resulting from interactions between micro- (niche), meso- (regime) and macro- (landscape) levels. Power is analysed as the capacity or incapacity of actors to mobilise resources and institutions to achieve a goal. This chapter concludes with a decolonial and Transmodern reading of the MLP, useful for understanding change processes in postcolonial countries.

Chapter 9 engages with diverse justice types (e.g., procedural, ecological, intergenerational) as part of seed systems change. It puts special emphasis on epistemic justice, as tensions between knowers and recognition of knowledge(s) are at the crux of the process. Andean seed systems change is framed here as a just transition, since historically marginalised actors mobilised power to transform the unjust structures of Modernity/Coloniality. By surfacing their knowledge and creating micro-institutions to strengthen their practices, SGNs have effectively materialised a fairer system.

The book concludes with a proposal of Transmodern transitions to *Buen(os) Vivir(es)*. These transitions begin with multiple actors converging around narratives of pluriversal biocentric futures and enacting them through shared practices. These articulations produce Transmodern innovative spaces strong enough to challenge established ways of doing things. Eventually, they open translational contact zones that enable dialogues of knowledge(s) with incumbent actors. These processes are premised on epistemic justice, expanding parameters of purposefulness and intelligibility of knowledge. As such, they allow for transformations in Transmodern terms, closing the abyssal line, and allowing the Modern and non-Modern to further integrate through Transmodern micro-institutions.

Ultimately, this book tells the story of civil society organisations overcoming Modernity/Coloniality to become influential players in seed systems. In this story, native seeds open translational contact zones that render visible that which has been made invisible by the colonial power matrix. These change processes evidence contemporary decolonial struggles and provide templates for Transmodern pluriversal futures to emerge. The book emphasises the importance of designing, nurturing, and shielding Transmodern micro-institutions. These spaces dare imagine and effectively implement ways of being otherwise. As such, more resources should be directed to Transmodern experiments that engage in organised transitions governance. Guided by radical visions of more sustainable and just food futures, Andean actors have come together to re-design agri-food systems. This book seeks to celebrate their efforts to decolonise such a critical system for the sustainment of human life on earth.

Notes

1 As opposed to universal. This concept positions plurality as a necessary condition for the future. It comes from the Indigenous Zapatista dictum of 'a world where many worlds fit'.
2 The United Nations Declaration on the Rights of Peasants and Other People Working in Rural Areas (p. 4), defines them as 'any person who engages or who seeks to engage alone, or in association with others or as a community, in small-scale agricultural production for subsistence and/or for the market, and who relies significantly, though not necessarily exclusively, on family or household labour and other non-monetized ways of organizing labour, and who has a special dependency on and attachment to the land'.
3 Family farming has been defined by the FAO as 'a way of organizing agriculture, livestock, forestry, fishing, aquaculture, and pastoralism, which is managed and operated by a family, and above all, depends predominantly on family labour by both women and men. Family and farm are linked, co-evolve, and combine economic, environmental, social, and cultural functions'.[1]

Reference

[1] S. Salcedo and L. Guzmán, Eds., *Agricultura familiar en América Latina y el Caribe: Recomendaciones de política*, 2014, p. 26. Accessed: 28 Jan. 2026. [Online]. Available: https://www.fao.org/4/i3788s/i3788s.pdf.

Part 1

Modernity/Coloniality and Transmodernity

1 Modern/Colonial seed systems

Modernity/Coloniality is now a globalised behemoth so naturalised that it is effectively invisible. This chapter draws from Latin American decolonial theories, post-Development, and other related fields to make visible the continuing colonial systems of power enforced since 1492.[1] The chapter defines Coloniality and focuses on three interrelated aspects: (i) Coloniality as inextricably intertwined with the Anthropocene event; (ii) climate crisis as a product of the material intensity of Modern/Western(ised) civilisation; and (iii) Modernity/Coloniality as an epistemic paradigm and an overlapping system of multiple power structures.

Probing deeper into the epistemic dimensions of Coloniality, the chapter mobilises the concept of the abyssal line to better understand the epistemic invention of colonial difference between a superior subspecies and the dispossessed, a dynamic that has framed multiple non-Western plural ways of being as obstacles to human progress. Modernity/Coloniality effectively rendered most of the world invisible and unpronounceable, creating an Anthropocene and an Anthropo-not-seen [1]. This chapter attempts to unveil the invisible side of Modernity by exploring its complexity and how it manifests in contemporary seed systems, particularly how an abyssal line persists in dividing the Modern and non-Modern, leading to fractures between formal and informal(ised) seed sub-systems.

Coloniality: The invisible side of Modernity

For Latin American decolonial scholars, Modernity is mainly a set of self-serving narratives [2]. These stories have told us that humans are superior to nature, that endless accumulation is possible through its subjugation, and that material growth equals progress [3]. This overarching myth of a standard civilisation originated in Europe some 500 years ago, when Iberian peoples invaded the continent now called America.

This event inaugurated Modernity as the first known fully global world system, which (despite constant attempts to conceal it) was colonial from the start [4]. Three interrelated aspects of Modernity as colonialism are especially relevant to sustainable transformations. First, colonialism can be considered

DOI: 10.4324/9781003715092-3

as inextricably intertwined with the Anthropocene, a 'time-transgressive geological event: the aggregated effects of human activities that are trans-forming the Earth system and altering biodiversity, producing a substantial record in sedimentary strata and in human-modified ground' [5, p. 350]. Some scholars understand the colonial invasions as interconnected to the Anthropocene due to two phenomena: the Columbian exchange (mixture of previously separate biotas) [6], and the Orbis spike (a dip in global atmospheric CO_2 due to human population decline in the Americas resulting from European diseases, war, enslavement, and famine), although an event framing of the Anthropocene aims to free the concept from alignment with established time units [5]. The Orbis spike in 1610 is important to the concept, because it connects the Anthropocene to issues of inequality in power relations between peoples, nations, and global economic growth [7].

Second, inequality in economic growth is inextricably tied to the colonial process. It is not coincidental that past imperial powers are rich while most past colonies are poor, as colonialism produced a global division of labour that continues to this day [8]. Additionally, there is an expanding body of work pointing to the material intensity of rich nations as a direct cause of the environmental crisis humanity faces. For instance, a study on national responsibility for ecological breakdown and fair share use of global resources shows that

> high-income countries, which represent only 16 per cent of the world population, are responsible for 74 per cent of resource use in excess of fair shares and are therefore the primary drivers of global environmental degradation, representing a process of ecological colonisation [9, p. 346].

Furthermore, the top 10 per cent of earners in the United States of America (USA) appear to produce around 40 per cent of greenhouse gas emissions in the country, evidencing marked environmental impact inequality [10].

Doing a full review of the literature framing rich nations as producers of planetary ecological breakdown exceeds the purpose of this book. The main point to be made is that colonial inequality persists to this day, and it is not only racial, epistemic, economic, and material but multidimensional. Additionally, this evidences that the Western civilisatory model itself is in crisis and Western solutions might not be viable. As Viola Recasens so aptly puts it:

> The collective hallucination that allowed people to believe for decades that the 'American way of life' (with its brutal patterns of consumption and per capita pollution) could one day become universal seems to have dissipated, forcing us to accept an unavoidable reality: the development model of the richest countries is not generalisable in space, it's not even extendable in time [11, p. 56].

The third aspect of Modernity and colonialism to be explored relates to its power and epistemic dimensions. This exploration will be conducted through the concept of Modernity/Coloniality,[2] which refers to the realisation of Modernity and colonialism as indivisible. Fifteenth-century colonialism ordered the world from a conception of European superiority, establishing first and foremost a racist and patriarchal power system. For Quijano, there are two foundational ideas to the story of Eurocentred Modernity: (i) human civilisation has a linear trajectory from the natural/primitive to Europe, making all non-Europeans pre-European and thus inferior; and (ii) differences between Europe and non-Europe are of (racial) nature and not the result of a history of power [4].

Modernity is the continuation of these, and other, systems of power, management, and control constituted by colonialism [2], [3]. These power structures created institutions in interaction with, and through, domination and exploitation of non-Europeans [8]. This was achieved through extraction of non-European cultural discoveries useful to the colonisers, repression of knowledge production, and forcing the colonised to partially learn the dominant culture to reproduce domination [4]. This process led to a duality-based otherisation that framed multiple peoples as an absolute deviant Other. Throughout history, multiple attempts have been made to save these inferiors, whether by being forced to convert to Christianity, correcting their backwardness via a civilising mission, aligning with the developmentalist project, or being further homogenised by the globalisation of neoliberalism [8], [12], [13], [14].

No matter the strategy used, Modernity has always been a project of Europeanisation [4]. However, Western Europe was only understood as the protagonist and sole producer of modernity until the 20th century, when the USA positioned itself as the endpoint of a linear history [2]. Whichever society is seen as the beacon of progress, this Western cultural expansion was never solely economic and political but epistemic in essence [2]. It positioned 'all other cultures as primitive, pre-modern, traditional and underdeveloped' [15, p. 39]; as living in an infantile stage of history.

This civilising project is described by decolonial scholars as the 'layered intersection of anthropocentric, androcentric, heterosexist, rationalist, Euro/Western-centric, modern/colonial, racialised, industrialist/developmentalist, capitalist and ableist systems of power' [3, p. 438]. Western knowledge created these systems by concealing its own bodily and geographical epistemic location, assuming a 'universalistic, neutral, [and] objective point of view' [8, para. 7].

This resulted in multiple asymmetrical core–periphery arrangements that have marginalised all those on the borders of these overlapping systems of power [3], [13]. These peripheral Others are the 'Global South(s)', a concept that signals onto-epistemic rather than geographical difference. In this sense, there are also Global Souths inside the borders of former imperial rich nations (e.g., Indigenous peoples, women, trans people, migrants, etc.).

Development as Coloniality

Modern/Colonial systems of power did not stop with the end of colonial rule [2], [12]; they are still perpetuated not only by their core but by the

Westernised elites of postcolonial countries [3]. This is tied to the reinvigoration of Modernity/Coloniality as Development, a discourse which defined most of humanity as 'underdeveloped'.

Development discourse emerged after World War II as Western Europe and its colonial administrations were in crisis. It was at this moment that the USA positioned itself and other industrialised nations at the apex of civilisation [14], [16]. In this way, Development discourse perpetuated a unilinear understanding of history with industrialism as its inevitable destiny, while excluding all other forms of social life [14].

Post-development scholars describe the ideology of Development as a way of keeping postcolonial countries from joining the communist system in the 20th century, rather affiliating them to the capitalist logic of market privileging and perpetual accumulation. This move maintained the international division of labour borne by Modernity/Coloniality (i.e., poor nations producing petty commodities for industrial production in rich nations) [17]. Additionally, this 'developed' way of life furthered inequality, cultural homogenisation, and detrimental ecological impacts that are currently too salient to ignore [14], [17]. As a manifestation of the invisible side of Modernity, Development discourse can be framed as the origin of the multiple global crises we presently face [3].

Development is now understood as a scientific form of knowledge [8], and has become diluted and naturalised as a force of social change that seems to be unquestionable [17]. This technocratic and managerialist intervention in the lives of the planet's majority perpetuates the hierarchisation of knowledge initiated by colonialism [16], [17]. Development discourse sustains the idea of deviance from a universal Eurocentric norm [17] and the globalisation of Westernisation [14]. As such, the Development model can also be understood as a 'mental, cultural and historical construct that has colonised the … world' [18, p. 305].

Even post-Modernity has been unable to transcend Coloniality, as it continues the universalising mission of Western being by ignoring the meta-narratives of the non-Modern[3] [2]. Interculturality as truly horizontal dialogue may be impossible in postmodernity, since relations of difference are still asymmetrical, and otherness is only allowed access or is integrated to *the* hegemonic frame of reference [19]. However, there are other options (See Chapter 2), since even after five centuries of Modern/Colonial hegemony, this has been an insufficient timeframe to eliminate or fundamentally transform all othered cultures that have survived in the peripheries [15].

The abyssal line[4]

Santos [20] developed the concept of the abyssal line to illustrate the exclusion produced by Modernity/Coloniality, which effectively rendered all otherised beings invisible or non-existent. Santos refers to this invisibility as produced by an epistemology of blindness and calls for an epistemology of seeing that allows for a constellation of knowledge(s) to exist. This section will deal with Santos's critical theory, expanding on the concept of the abyssal line and abyssal thinking.

For Santos, colonialism produced an epistemic invention of difference that divided the world between a superior subspecies (Western man) and masses of dispossessed (the Global Souths). Through centuries of disqualification of Global South agents as ignorant, inferior, local, particular, backward, unproductive, and lazy, they have become invisible, inaudible, and illegible to the core of Modern/Colonial power. This disqualification has effectively resulted in an epistemicide,[5] through tactics like the destruction of social practices, epistemic control,[6] appropriation (incorporation, co-optation, and assimilation), and violence (physical, material, cultural, and human destruction).

For Santos, an 'abyssal line' was produced and is maintained by Modernity/Coloniality. This line divides colonial societies from metropolitan ones even decades after the end of colonialism. Furthermore, it divides social reality itself in such a way that 'whatever lies on the other side of the line remains invisible or utterly irrelevant' [20, Ch. 2]. Santos thus understands Modern Western thinking as abyssal thinking, a 'system of visible and invisible distinctions, [where] the invisible ones [are] the foundation of the visible ones' [20, Ch. 4]. Abyssal thinking makes the co-presence of both sides of the 'abyssal line' impossible, effectively eliminating whatever realities are on the other side.

One of the key consequences of this Global North/Global South divide is that the Global North seems unable to learn in non-colonial terms. This is critical, because there appears to be an exhaustion of Western thought and its critical theories facing the multiple challenges of the Anthropocene. This does not mean that the Global Souths hold all the answers, as evidenced by myriad failed decolonial and post-capitalist political projects. It is for this reason that Global North–Global South horizontal dialogues are essential, for which the concepts of the abyssal line and abyssal thinking are important.

The abyssal line is the product of epistemological monopolisation, a process that rendered some knowledge(s) incommensurable and incomprehensible to the Modern/Colonial core. Modern institutions like capitalism, law, and scientific knowledge consolidated through this process, and now have a monopoly determining what is true and false. Accordingly, the abyssal line manifests in multiple social spaces, leading to conceptualisations of diverse lines. Of special interest to this book are the main lines identified by Santos [20]: an abyssal epistemological line, and lines in law and science.

These lines intersect with contemporary seed systems in complex ways, as elaborated in the next section. Epistemologically, Modern and non-Modern seed knowledge(s) are different and constantly engage in dialogue, creative tension, and antagonism. The legal abyssal line leaves vast spaces of social reality as lawless, by keeping them in a non-legal space. Finally, the scientific line makes invisible all non-scientific technological development, while Global South social movements understand scientific knowledge as just one knowledge amongst many.[7]

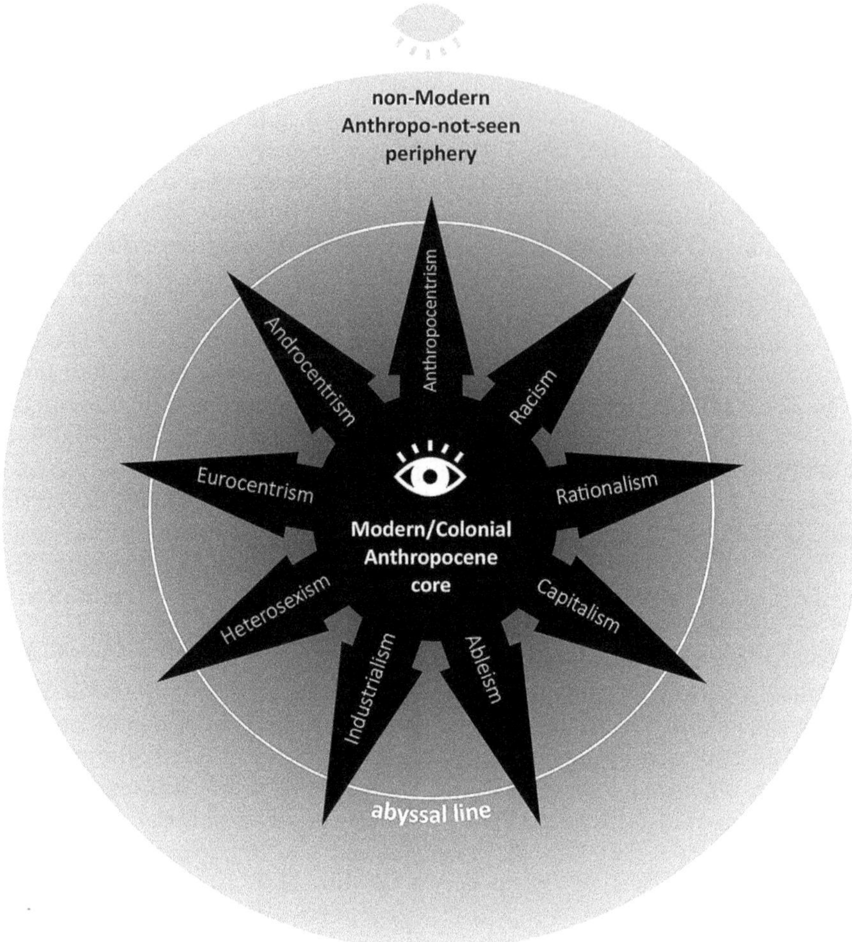

Figure 1.1 The overlapping systems of power of Modernity/Coloniality producing the abyssal line

Coloniality in seed systems

This section aims to show that the centuries-long process of agricultural industrialisation is tied to the dynamics of Coloniality and abyssal thinking. It argues that it is ultimately an effort to Westernise and standardise plural food practices and narratives, going as far as homogenising seeds as stable commodities. The abyssal line produced by colonialism, modernisation, and Development is patently clear in seed systems. At the core of these systems there is a formal way of doing things and relating to seeds where technoscience, capitalism, and corporations play a central role. On the peripheries, there are plural ways of relating to food and doing agriculture that respond to ancestral and traditional knowledge(s). This 'informal' seed system is generally seen by incumbents in the system as an obstacle to progress, since it has been marginalised due to epistemic, scientific, and legal abyssal lines.

A full review of the history of seed systems and their industrialisation is beyond the scope of this book. However, two important aspects of this history are included below to provide context to the transformation processes described in subsequent chapters. First, the concept of food regimes provides an analysis of food system dynamics at global scale, allowing for the identification of stable food regime moments within Modernity. This periodisation offers insights into the articulation of capital and states in structuring food systems. Second, and as part of these structuring processes, the green revolution is described as the process by which industrial agriculture was globalised and stablished as the dominant configuration for food systems. Reviewing these two aspects of food history will hopefully evidence how the abyssal line crosses food systems at global scale.

A brief history of Modern seed systems

Food regimes

Food Regime Theory seeks to understand 'the organisation of food production, distribution, and consumption on a world scale in political economy' [21, p. 231]. Food regimes are used as a method of analysis to 'identify significant relationships and contradictions in capitalist processes across time and space' [22, p. 163]. This analysis has led scholars to identify a series of historical periods with distinct rules, and modes of production and consumption.

These periods are not strictly bounded but refer more to a series of continuous changes in global structuring processes particular to the history of capitalism [22]. Food regimes describe a configuration of food relations at global scale that support and stabilise capitalist growth [21]. This ties to the emergence and stabilisation of industrial agriculture, which is linked to processes of mechanisation and chemicalisation and, more broadly, to a neoliberalisation of nature where genetics (e.g., seeds, animals) are framed as corporate private property [23].

Food regime characterisation has been the subject of debates for over 30 years. Ongoing debates mean that periodisation is still contested [24]. Polemics relate to the number of regimes, their characteristics, or their timeframes. Generally, there is some agreement on three food regimes (See Table 1.1): the first from 1870 to 1914–1930, the second from the 1940s to the 1970s, and a third starting from the late 1980s [22], [23]. However, this periodisation has been criticised on several fronts. For instance, population variables are unaccounted for, and peasant farmers are absent in descriptions of the first and second regimes. Additionally, debates regarding the third regime seem to have shifted towards demonstrating downsides of agribusiness while highlighting the virtues of small-scale farming [23].

Table 1.1 Three food regimes

Regime	*Timeframe*	*Description*
First	1870–1914/1930	Food provision to the emergent European industrial class from tropical imports and grain from settler colonies.
Second	1940s–1970s	Redirection of food surplus from the USA to poor nations as part of Development. Motivated by Cold War fears and aimed at integration to Western markets while avoiding implementations of communism. This leads to some nation-building and industrialisation in poor postcolonial nations while creating another agricultural international division of labour.
Third	1980s–present	Amplification and consolidation of international supply chains leading to displacement of small-scale farmers to peripheries of cities. Emergence of a global food–fuel complex and expansion of the supermarket model. Alternatives emerge to resist this regime in the form of social movements promoting food sovereignty, slow food, organic farming, and the peasant way, among others. Accordingly, competing narratives of extractive and regenerative food systems emerge and contest food futures [25].

Four additional problems in food regime characterisations have been identified: first, causal dynamics underlying the first regime; second, assumptions of continued hegemony of the USA from the 19th to the 20th century related to nation-building in poor nations through food aid/exports; third, temporal delimitations of international regimes; and fourth, an assumption that settler agriculture was fundamental to the formation of nation-states [21]. This has led to a proposal of new food regimes: a 'first national capitalist food regime' (1750–1846), a 'first international or liberal food regime' (1846–1870), a 'second international or imperial food regime' (1870–1930), a 'third international or political productivist food regime' (1930–1980), a 'fourth international or neoliberal food regime' (1980–2010), and a 'fifth international or post-neoliberal food regime' [21].

The following section zooms in on the second food regime to explore in more detail the globalisation of industrial food production. It provides an overview of the green revolution, understanding it as a series of ongoing agricultural change processes [26], [27]. In this sense, it is a long green revolution that begins with state support, investment, and philanthropy (from the 1940s), giving way to a financialisation of agriculture led by the private sector (ongoing as part of the third regime). The mythos of the green revolution is one of technological triumph over hunger while it reconfigured states and society for better or worse [27]. The following section attempts to briefly explore the green revolution as it relates to seeds, particularly the ways in which it was implemented globally, its shortcomings, and how it emerged from and perpetuated an abyssal line in seed systems.

The green revolution

Since the late 19th century, seeds have transitioned from being a deeply rooted commons and a public good to an increasingly privatised commodity. This has been mainly enabled by advances in plant breeding and genetics [28], [29], [30], as well as an encroachment of intellectual property rights (IPR) over germplasm [31]. An important moment of acceleration came after World War II, when European agriculture fully industrialised to achieve food security in a time of high uncertainty [32]. The green revolution that soon followed established a global design of agricultural development for Third World nations based on extensive monocultures that needed improved seeds, sophisticated machinery, and large amounts of agrichemicals [33], [34].

The green revolution was kickstarted by the development of high-yield varieties (HYVs) as part of crop improvement programmes. These initiatives were tied to plant-breeding institutions in poor nations funded by donors and aid agencies from rich nations. This process was initiated by the USA as an agricultural development programme aimed at Latin America in general and Mexico in particular. It is important to highlight that the development of HYVs involved the collection of indigenous germplasm without compensating the communities or countries of origin. This genetic material was extracted and used to create new varieties to be sold to origin countries, often leading to displacement or disappearance of the native or landrace seed varieties used by breeders in rich nations [35].

Crop improvement programmes and HYV development were partly motivated by perceived obligations to new postcolonial countries in Africa and Asia, opportunities for new markets, and Cold War power dynamics (i.e., favouring a green revolution over a red revolution) [27], [36]. Although ideologically framed as aiding national development, 'the reality was an internationalisation of agribusiness chains of inputs, technologies and foodstuffs [that eroded] the coherence of national farm sectors' [22, p. 146] in poor countries.

Through the green revolution, developing countries were effectively discouraged from investing in local agricultural systems, privileging food aid over indigenous agriculture as a solution to famine and malnutrition [37], [38]. When investments occurred in poor nations, the green revolution mainly benefited large producers by deploying a particular type of technological package used in large-scale agriculture in the USA. Despite being generally seen as scale-neutral (i.e., also adopted by smallholders due to self-interest), the green revolution exported knowledge and technology useful to farmers with large states and a commercial disposition [27]. As such, the green revolution can simultaneously be framed as 'processes of state reconfiguration, capitalist accumulation, concentration of power, disenfranchisement, agricultural investment and innovation' [27, p. 2].

The green revolution would not have been possible without governments; on the one hand, through research by adapting HYVs to local agro-ecologies

[36]; and on the other, as providers of incentives, subsidies, or infrastructure (e.g., irrigation systems) [39]. The green revolution would not have succeeded without state subsidies. However, governments were always constrained by capitalism's tenets of private property and individual free enterprise, which led them to frame the problem of hunger in particular ways [27]. These capitalist underpinnings became more salient as the third food regime emerged and the World Bank displaced states to become the principal funder of the green revolution. This new configuration included small-scale farmers in lending schemes to counteract criticisms of only benefiting big commercial farming [27].

The green revolution has certainly increased yield and quality for the few crops subject to scientific intervention, as well as per capita income in developing countries [26], [29], [36]. However, there has been a simultaneous decay of genetic diversity, leading to increased crop homogeneity and reduced resilience [32], [40]. Some contend that it has undermined human nutrition, as crops rich in micro-nutrients (generally produced by peasants) are replaced by cereal grains with lower nutritional value [27]. Additionally, the increase in production promised by the green revolution appears to have stagnated and has led to soil degradation, decreased water quality, and other environmental issues [33]. Impacts are not only felt in poor nations but also perceived as threats to human health in green revolution originators like the USA [24].

While malnutrition persists, the green revolution keeps benefiting elites while increasing inequality between rich and poor farmers [39]. This relates to the creation of a class of landless poor due to incentives for international buyers to participate in land markets in emerging economies. These land-grabs, as well as other green revolution dynamics, disproportionately affect women, leading to additional questions about gender imbalances [27].

Adding to these impacts, the last 30 years have seen increased financialisation of food via agricultural derivatives [32], and alarming levels of concentration in the seed industry [29], [33], [41]. Although concentration in food markets has occurred for more than one hundred years and at many levels [42], ten companies now control around two-thirds of global proprietary seed [30]. In recent years, the global seed market went from being dominated by six companies[8] [40] to four: Bayer, Corteva, ChemChina-Syngenta, and BASF [43]. Crucially, this concentration may be more pronounced at sub-global level and for specific products (e.g., maize in Brazil) [42]. This phenomenon is enabled by legislative framings of seeds as private property, privileging private breeder rights over other system actors (e.g., smallholder farmers, and public and even small private plant breeders) [41], [44]. These framings, in turn, respond and are reinforced by international trade and investment law, which has been used to globalise industrialist agriculture, setting up a multilateral trading system instead of ensuring food security or sustainability [38].

Global seed regulation responds more to a collection of legislative packages than to any individual source [32]. These laws include intellectual property rights (IPR) on plant material, plant breeder rights (PBR), phytosanitary and biosecurity norms, commercialisation and quality standards, and registration and certification schemes [30], [32], [33], [34]. Regarding IPR, the International Union for the Protection of New Varieties of Plants (UPOV) plays a pivotal role. This intergovernmental organisation was established by the International Convention for the Protection of New Varieties of Plants in 1961, with revisions in 1972, 1978, and 1991. Further reinforcing control over plant material, the World Trade Organization's Trade-Related Intellectual Property Rights agreement (TRIPS) expanded intellectual property protection for plant varieties in 1994 [29].

In the later scheme of UPOV, breeders can obtain IPR over plants in the form of plant variety rights (PVR) by fulfilling the criteria of distinctiveness, uniformity, and stability.[9] UPOV standardisation led to innovation in the sector but further marginalised non-industrial seeds [32], also excluding many farmer-bred or organic cultivars [41]. Past versions of the UPOV convention allow some freedom to multiply and use proprietary seeds.[10] However, the 1991 convention grants breeders exclusive rights to protected varieties, effectively prohibiting customary seed exchanges amongst farmers [29], [30]. UPOV now has 77 members: 17 states bound to the 1978 act, and 58 states and two organisations affiliated to the 1991 act.[11]

These supranational agreements inform seed policy at a national level, generally promoting the standardisation of seeds [45]. However, the multiscalar nature of seed regulation leads to contradictions at the national level when responding to supranational agreements.[12] One dimension of this issue is the tension between national sovereignty over genetic resources and supranational pressures to conform life forms to IPR standards [30]. These legislations also clash with traditional community-based seed systems [29], eroding their capacity to save seeds for resowing and effectively diminishing farmer sovereignty over seeds [30], [41]. This has not only affected smallholder farmers but also public sector breeders, aiding the process of concentration by large-scale transnational corporations [41].

These legislative pressures over seed systems are deeply impactful in Latin America, as AfroIndopeasant,[13] family, and small-scale agricultural practices have always been an important part of food systems. In fact, over half of food in the region may be produced by smallholders participating in non-industrial practices [30], [33]. Despite their importance for national food security, this subsystem has been constantly marginalised and perceived as an obstacle on the road to modernisation. Indigenous, small-scale, peasant, and family agriculture actors are generally excluded from decision-making in the region, while relevant governance spaces are almost exclusively composed of incumbents with an industrialist bias (be they representatives from government, academia, or industry).

This governance model has produced divisions and deep mistrust between two seed sub-systems, generally referred to as the formal and informal seed systems. Informal system actors generally feel policies benefit big transnational producers without thinking of wider impacts, while formal system actors see the informal system as stagnant and unwilling to modernise.

Seeing the abyssal line in seed systems

This section deepens the exploration of the formal and informal seed sub-systems and their relation to Modernity/Coloniality. Scholars tend to distinguish between two seed systems that are defined in different ways but share similar contours (see Table 1.2). These systems intertwine, often overlap, and constantly adapt to each other's pressures. However, these pressures are asymmetrical for two main reasons: (i) growth of transnational corporations depends on the erosion of informal seed systems to expand their markets; and (ii) there is a state-level agro-industrial bias, a product of a perceived superiority of the formal system, supported by the aforementioned seed laws that curtail possibilities for informal systems [30], [33], [44].

Despite the overlaps between these two sub-systems, there are distinct seed types in each camp: (i) certified or improved seeds in the formal system; and (ii) native, creole, landrace, heirloom, or local seeds in the informal system. Improved seeds are hybridised to acquire desirable traits and must be purchased seasonally to produce the same variety [40]. Farmers license the use of these seeds as part of a rental agreement of sorts that prohibits the saving and reuse of germplasm [31], [41]. These seeds respond to DUS criteria (distinctiveness, uniformity, and stability) and are certified by national food authorities.

Table 1.2 Different denominations of seed systems

Author	*Name*	*Description*
Felicien et al. [33]	Modern	
	Traditional	
Wattnem [30]	Formal	Regulated scientific plant breeding with public and private participation.
	Informal	Farmer-led unregulated or loosely regulated.
García López et al. [45]	Industrial	'large-scale production and supply of commercial seed varieties … [subject to] … strict quality control based on standard physical and physiological criteria' [45, p. 829].
	Local	Dependent on seed-saving practices and trust-based community exchanges sustained by peasant, Indigenous communities, and small or medium-scale farmers.
Bernstein [46]	Global capitalist agriculture	Large-scale industrial farming.
	Peasant mode of production	Small-scale customary farming.

Regarding the informal system, native seeds originate from the same place where they are sown. Creole seeds may not come from the same place but have been adapted by farmers to local conditions [34].[14] A landrace or peasant variety has not been subject to formal improvement, but has been transformed by the traditional knowledge of peoples directly involved in growing it [32]. And heirloom varieties have been historically saved and maintained over long periods of time by people [40].

A narrative has consolidated that frames industrial or transgenic seeds as unique or improved, while all others are understood as basic, conventional, or simply as raw genetic material [47], [48]. Furthermore, seeds in the informal system are considered uncertifiable, since they are associated with low quality and productivity, as well as spreading disease [34]. These imaginaries seem to negate the fact that collectively produced informal seeds have been necessary throughout history for all genetic improvement programmes [33]. The formal system has always been enabled by informal system germplasm [30], often through biopiracy, while farmers are simultaneously persecuted by national authorities for breaching corporate seed packet licences [41]. Biopiracy refers to the appropriation of native seeds and associated Indigenous/traditional knowledge without the consent of peoples involved in their development [29], [34], [46], [47].

The formal and informal seed sub-systems respond to more than distinct seed types, technologies, knowledge(s), or production practices. Their differences point to often contradictory worldviews. To exemplify this, in the case of India, Shiva sees two paradigms of biodiversity at play: one, a destructive genetic extractivism that disregards the needs of poor farmers; the other, a recognition of life's interdependence in its role supporting farmers' livelihoods [47]. In the case of Colombia, Vidal and Escobar see conflicts between two forms of understanding seeds: one where seeds become the capital of the agribiotech industry and knowledge associated with their cultivation and preservation is privatised; and another where seed, territory, and knowledge are interwoven to conceive and use seeds as a common good [34].

This book sets out to explore these contrasting systems as inextricably tied to Modernity/Coloniality. The often-opposing narratives and practices in the global seed system emerge from profound differences in worldviews or *what is* our reality (ontology) and *how we know or understand what is* (epistemology). Analysing 21st century Andean seed system transformations evidences the tensions between the Modern and non-Modern and their diverging concepts of progress or what it means to live a good life. Furthermore, this analysis raises questions about power, justice, and what constitutes valid knowledge. It begs to move beyond formal and informal categories by exploring the profound differences of the Global South(s) and the deep-seated assumptions naturalised by Coloniality, the

> complex and multidimensional legacy of divisive, exploitative, stratifying and hierarchical forms of power (e.g., Eurocentric/Western-centric hegemony), forms of knowledge (e.g., technoscientific instrumental rationality), forms of (inter)subjectivity (e.g., possessive individualism), forms of human interrelations (e.g., racism, classism, heteropatriarchalism, etc.), and forms of human dominion over land and mastery of 'nature' (e.g., anthropocentric property/dominion/sovereignty) that have become entrenched and continue to be reproduced throughout the world as an ongoing consequence of colonisation [3, p. 432].

It can be argued that divisions between formal and informal seed systems are the product of the abyssal line. Modern institutions like science, law, and capitalism reinforce the formal system while ignoring (thus informalising) other seed narratives, practices, rules, and technologies. Contemporary seeds appear to be valuable only when big agricultural corporations subject them to technoscientific intervention to commodify them. By contrast, all the knowledge necessary to maintain native agrobiodiversity is made invisible, as governments, corporations, universities, and other powerful actors tend to see native seeds as raw, naturally occurring materials. Consequently, seeds appear to be only considered a technology when they are modified in a lab by scientists, and not when they are modified in the field by Afro-Latino, Indigenous, or peasant farmers.

An abyssal legal line is also evident in seed systems, as detailed legislative scaffoldings for the formal system are commonplace, while rules are scarce for the informal system. Formal system rules have transnational, national, and local effects, protecting the IPR of commodified seeds while conveniently ignoring the biopiracy of native seeds and their associated knowledge. This robust set of rules effectively disregards the informal seed system, keeping it in an a-legal space where it is simultaneously policed and controlled without being fully recognised. As Ana Felicien puts it, there are no informal seed systems, just informalised sub-systems [49]. Despite these shortcomings in the field of law, this book will show that change has occurred and there is a move towards more inclusive legal frameworks. These transformations have been spearheaded by social movements proposing alternatives within the emergent third food regime.

Seed guardian networks cutting through the abyssal line

From an understanding of the green revolution as a success (completed and unequivocally effective), advocates have called for a second green revolution. This renewed process of agricultural modernisation would attend to the criticisms of the first revolution by being more inclusive of smallholders and women, while considering environmental impacts. However, it maintains a framing of small-scale and AfroIndopeasant ways of producing food as unsustainable and backwards, particularly in Africa where it is meant to have the biggest impact [27]. As the third food regime consolidates, it is increasingly evident that social movements around the world are proposing alternatives beyond the green revolution. Trans-local articulations hybridising narratives of food sovereignty, vernacular knowledge, and technologies, and alternative food practices are providing a counterweight to more of the same Modern models.

This book focuses on some of these trans-local food social movements operating in three countries of the Latin American Andes, particularly seed guardians, who emerged as a response to the intensification of pro-industry seed legislation, which undermined existing non-industrial practices. Seed guardian networks (SGNs) have consolidated over the last 25 years. These networks defend the rights of peoples to produce, store, improve, exchange, and trade their seeds according to Indigenous, Afro-descendant, peasant, agroecological, and organic agricultural narratives and practices. They can be described as decentralised, place-based, trans-local coalitions of people and organisations committed to safeguarding landrace, native and creole seeds, and their associated practices of care.

SGNs understand seeds as the starting point of nourishment, but also as free, living entities that embody the origin and regeneration of existence. For them, humans have a deep relationship of nurturing and care with seeds, a connection that mobilises narratives of seeds as being part of family and community. In this space, plural nurturing practices have led to a wide diversity of agricultural seeds influenced by generations of localised knowledge(s), experiences, cultures, and traditions. Seeds are also understood as an intergenerational commons beyond government and corporate control, sometimes going as far as framing seeds as entities with agency and rights of their own.

SGNs have become influential players in Latin American seed systems, as they simultaneously resist hegemony and propose alternatives. They emerged as a response to the intensification of neoliberal policies that opened spaces for transnational corporations while curtailing possibilities for non-industrial practices. Initially, the activities of SGNs were mostly directed towards resisting legislative changes and getting their voices heard. Over time, they have been able to consolidate, effectively creating protected spaces for non-hegemonic seed practices (see Chapter 3). SGNs can be considered champions of the informal (ised) side of seed systems. They promote amalgamations of AfroIndopeasant practices with agroecology and organic agriculture, revitalising non-Modern narratives and practices connected to local agrobiodiversity.

Throughout this book, three cases of seed system transformations in the Latin American Andes will be explored. Special attention will be given to SGNs and their role in bringing forth sustainable transformations. These cases evidence the presence of abyssal lines and show how historically marginalised food system actors have been able to transcend them. The unique transformation processes of Colombian, Ecuadorian, and Venezuelan seed systems show signs of a new kind of future emerging. In these processes, SGNs play a central role in making visible non-Modern seed practices and narratives. Through their efforts, state–society relations shift, opening spaces for hybrid non-Modern and Modern institutions.

The philosophical underpinnings of these emerging (non-Modern ∩ Modern) hybrids will be explored in the next chapter, seen through the lens of two emerging intellectual projects: Transmodernity and *Buen Vivir*.

Notes

1 1492 marks the invasion of the continent now named America by Iberian peoples.
2 'Coloniality' was initially conceptualised by Aníbal Quijano [4] and is used interchangeably with 'Coloniality of power' and the 'colonial matrix of power'.
3 Non-Modern denotes the 'coexistence of temporalities and modes of living and thinking that are neither premodern nor postmodern' [2, p. 117].
4 This section draws solely from the work of Boaventura de Sousa Santos and his rearguard theory developed in *Epistemologies of the South* [20].
5 Epistemicide is the systematic murder of knowledge(s).
6 Controlling othered ways of knowing, for example, coercively limiting the use of Indigenous languages.
7 For an example of these politics of knowledge at play in India see [50].
8 Monsanto, DuPont, Syngenta, Bayer, Dow, and BASF.
9 Also known as the DUS criteria.
10 The 1978 version allows the use of protected varieties by breeders and seed saving (the practice of keeping reproductive material to resow in future seasons) by farmers.
11 For a specific list, visit: www.upov.int/edocs/pubdocs/en/upov_pub_437.pdf.
12 For example, in the case of Ecuador see [32], and for Colombia see [28], [34], [51].
13 Throughout the book, this term will refer to Afro-Latino, Indigenous, and peasant communities. While recognising their diversity, this term emphasises their commonalities, particularly as they relate to food practices.
14 Creole seeds also include adapted certified seeds.

References

[1] M. De la Cadena, 'Uncommoning nature', *E-flux*, pp. 1–8, 2015. [Online]. Available: http://supercommunity.e-flux.com/texts/uncommoning-nature/.

[2] W. Mignolo and C. Walsh, *On decoloniality*. Durham, NC: Duke University Press, 2020. doi:10.1515/9780822371779.

[3] L. E. Figueroa Helland and T. Lindgren, 'What goes around comes around: From the Coloniality of power to the crisis of civilization', *Journal of World-Systems Research*, vol. 22, no. 2, pp. 430–462, 2015, doi:10.5195/JWSR.1.

[4] A. Quijano, 'Colonialidad del poder, eurocentrismo y América Latina', in *Cuestiones y horizontes: De la dependencia histórico-estructural a la colonialidad/descolonialidad del poder*. Buenos Aires: CLACSO, 2014, pp. 777–832. [Online]. Available: http://biblioteca.clacso.edu.ar/clacso/se/20140507042402/eje3-8.pdf.

[5] P. L. Gibbard *et al.*, 'A practical solution: The Anthropocene is a geological event, not a formal epoch', *Episodes*, vol. 45, no. 4, pp. 349–357, Dec.2022, doi:10.18814/epiiugs/2021/021029.

[6] A. W. Crosby, *The Columbian exchange: Biological and cultural consequences of 1492*. Westport, CT: Praeger, 2003.

[7] S. L. Lewis and M. A. Maslin, 'Defining the Anthropocene', *Nature*, vol. 519, no. 7542, pp. 171–180, 2015, doi:10.1038/nature14258.

[8] R. Grosfoguel, 'Decolonizing post-colonial studies and paradigms of political-economy', *Transmodernity*, vol. 1, no. 1, 2011.

[9] J. Hickel, D. W. O'Neill, A. L. Fanning, and H. Zoomkawala, 'National responsibility for ecological breakdown: A fair-shares assessment of resource use, 1970–2017', *Lancet Planet Health*, vol. 6, no. 4, pp. e342–e349, 2022, doi:10.1016/S2542-5196(22)00044-4.
[10] J. Starr, C. Nicolson, M. Ash, E. M. Markowitz, and D. Moran, 'Income-based U.S. household carbon footprints (1990–2019) offer new insights on emissions inequality and climate finance', *PLOS Climate*, vol. 2, no. 8, p. e0000190, Aug.2023, doi:10.1371/journal.pclm.0000190.
[11] A. V. Recasens, '"Pachamamista" discourses versus development policies: The debate over *Sumak Kawsay* in the Andes', *Revista de Ciencias Sociales*, vol. 48, pp. 55–72, 2014.
[12] R. Grosfoguel, 'Transmodernity, border thinking, and global Coloniality – Decolonizing political economy and postcolonial studies', *Eurozine*, pp. 1–23, 2008. [Online]. Available: www.eurozine.com/transmodernity-border-thinking-and-global-coloniality.
[13] W. Mignolo, *Trayectorias de re-existencia: ensayos en torno a lacolonialidad/decolonialidaddel saber, el sentir y el creer.* Bogota: Universidad Distrital Francisco José de Caldas, 2015.
[14] W. Sachs, *The development dictionary: A guide to knowledge as power*, 2nd ed., vol. 18, no. 1. London: Zed Books, 2010. doi:10.2307/1971880.
[15] E. Dussel, 'Transmodernity and interculturality: An interpretation from the perspective of philosophy of liberation', *TRANSMODERNITY: Journal of Peripheral Cultural Production of the Luso-Hispanic World*, vol. 1, no. 3, 2012.
[16] M. C. Misoczky, 'World visions in dispute in contemporary Latin America: Development x harmonic life', *Organization*, vol. 18, no. 3, pp. 345–363, 2011, doi:10.1177/1350508411398730.
[17] A. Ziai, 'Post-development 25 years after the development dictionary', *Third World Q*, vol. 38, no. 12, pp. 2547–2558, 2017, doi:10.1080/01436597.2017.1383853.
[18] A. E. Beling, J. Vanhulst, F. Demaria, V. Rabi, A. E. Carballo, and J. Pelenc, 'Discursive synergies for a "great transformation" towards sustainability: Pragmatic contributions to a necessary dialogue between human development, degrowth, and *Buen Vivir*', *Ecological Economics*, vol. 144, March, pp. 304–313, 2018, doi:10.1016/j.ecolecon.2017.08.025.
[19] D. Añazco, 'La interculturalidad como nuevo objeto: Esbozo de un planteamiento transmoderno', *Ciencia e Interculturalidad*, vol. 24, no. 01, pp. 106–119, 2019, doi:10.5377/rci.v24i01.8006..
[20] B. de S. Santos, *Epistemologies of the South – Justice against epistemicide.* New York: Routledge, 2016.
[21] M. Tilzey, 'Food regimes, capital, state, and class: Friedmann and McMichael revisited', *Sociol Ruralis*, vol. 59, no. 2, pp. 230–254, Apr.2019, doi:10.1111/soru.12237.
[22] P. McMichael, 'A food regime genealogy', *Journal of Peasant Studies*, vol. 36, no. 1, pp. 139–169, 2009, doi:10.1080/03066150902820354.
[23] H. Bernstein, 'Agrarian political economy and modern world capitalism: The contributions of food regime analysis', *Journal of Peasant Studies*, vol. 43, no. 3, pp. 611–647, May2016, doi:10.1080/03066150.2015.1101456.
[24] J. Beacham, 'Planetary food regimes: Understanding the entanglement between human and planetary health in the Anthropocene', *Geographical Journal*, vol. 188, no. 3, pp. 318–327, Sep.2022, doi:10.1111/geoj.12407.
[25] M. D. Anderson and M. Rivera-Ferre, 'Food system narratives to end hunger: Extractive versus regenerative', *Current Opinion in Environmental Sustainability*, vol. 49, pp. 18–25, doi:10.1016/j.cosust.2020.12.002.

[26] R. D. Hurt, '7 reflections', in *The green revolution in the Global South: Science, politics, and unintended consequences.* Tuscaloosa, AL: University of Alabama Press, 2020, doi:10.2307/jj.30297352.

[27] R. Patel, 'The long green revolution', *Journal of Peasant Studies*, vol. 40, no. 1, pp. 1–63, Jan.2013, doi:10.1080/03066150.2012.719224.

[28] L. Gutiérrez Escobar and E. Fitting, 'The red de Semillas libres: Contesting biohegemony in Colombia', *Journal of Agrarian Change*, vol. 16, no. 4, pp. 711–719, 2016, doi:10.1111/joac.12161.

[29] E. Senini, 'Farm saved seeds: A right to use or a right abused?' *European Food and Feed Law Review*, vol. 13, no. 2, pp. 116–124, 2018.

[30] T. Wattnem, 'Seed laws, certification and standardization: Outlawing informal seed systems in the Global South', *Journal of Peasant Studies*, vol. 43, no. 4, pp. 850–867, 2016, doi:10.1080/03066150.2015.1130702.

[31] C. H. Luby and I. L. Goldman, 'Freeing crop genetics through the Open Source Seed Initiative', *PLoS Biol*, vol. 14, no. 4, pp. 1–7, 2016, doi:10.1371/journal.pbio.1002441.

[32] C. Gevers, H. F. M. W. van Rijswick, and J. Swart, 'Peasant seeds in France: Fostering a more resilient agriculture', *Sustainability (Switzerland)*, vol. 11, no. 11, pp. 1–22, 2019, doi:10.3390/su11113014.

[33] A. Felicien, 'Historia de las leyes de semillas y la modernidad en la agricultura', in *Semillas del Pueblo*, M. Pérez et al., Eds. Caracas: Editorial La Estrella Roja, 2016, pp. 21–49.

[34] N. H. Vidal and L. G. Escobar, 'Epistemic and political struggles against the privatization of seeds and collective knowledges', *Revista Colombiana de Antropologia*, vol. 55, no. 2, pp. 39–63, 2019, doi:10.22380/2539472X.798.

[35] J. Kloppenburg, 'Seeds of struggle: Plant genetic resources in the world system', in *First the seed: The political economy of plant biotechnology*, 2nd ed., 2004. Madison, WI: University of Wisconsin Press.

[36] D. Gollin, C. W. Hansen, and A. M. Wingender, 'Two blades of grass: The impact of the green revolution', *Journal of Political Economy*, vol. 129, pp. 2344–2384, 2021, doi:10.1086/714444.

[37] E. Gilson and S. Kenehan, Eds., *Food, environment, and climate change justice at the intersections.* Lanham, MD: Rowman & Littlefield, 2021.

[38] S. Narula, 'Beyond reform: Food sovereignty and the future of global food systems', *Indiana Journal of Global Legal Studies*, vol. 31, no. 2, pp. 141–203, 2024. [Online]. Available: https://muse.jhu.edu/article/949739.

[39] R. D. Hurt, 'Introduction', in *The green revolution in the Global South.* Tuscaloosa, AL: University of Alabama Press, 2020, doi:10.2307/jj.30297352.

[40] T. Volkening, 'Seeds with stories: Seed savers exchange revisited', *Journal of Agricultural and Food Information*, vol. 19, no. 2, pp. 106–115, 2018, doi:10.1080/10496505.2018.1441578.

[41] J. Kloppenburg, 'Re-purposing the master's tools: The Open Source Seed Initiative and the struggle for seed sovereignty', *Journal of Peasant Studies*, vol. 41, no. 6, pp. 1225–1246, 2014, doi:10.1080/03066150.2013.875897.

[42] J. Clapp *et al.*, 'Corporate concentration and power matter for agency in food systems', *Food Policy*, vol. 134, 2025, doi:10.1016/j.foodpol.2025.102897.

[43] J. Clapp, 'The problem with growing corporate concentration and power in the global food system', *Nature Food*, vol. 2, no. 6, pp. 1–5, doi:10.1038/s43016-021-00297-7.

[44] A. Felicien *et al.*, 'Exploring the "grey areas" of state-society interaction in food sovereignty construction: The battle for Venezuela's seed law', *Journal of Peasant Studies*, vol. 47, no. 4, pp. 648–673, 2020, doi:10.1080/03066150.2018.1525363.

[45] V. García López, O. F. Giraldo, H. Morales, P. M. Rosset, and J. M. Duarte, 'Seed sovereignty and agroecological scaling: Two cases of seed recovery, conservation, and defense in Colombia', *Agroecology and Sustainable Food Systems*, vol. 43, no. 7–8, pp. 827–847, 2019, doi:10.1080/21683565.2019.1578720.

[46] H. Bernstein, 'Food sovereignty via the "peasant way": A sceptical view', *Journal of Peasant Studies*, vol. 41, no. 6, pp. 1031–1063, 2014, doi:10.1080/03066150.2013.852082.

[47] V. Shiva, 'Recovering biodiversity', *Soc Change*, vol. 31, no. 1–2, pp. 21–37, 2001, doi:10.1177/004908570103100204.

[48] D. Silva Garzon, 'Three logics of action and reaction for the monopolization of seed markets in Colombia', *Revista Colombiana de Antropologia*, vol. 55, no. 2, pp. 9–37, 2019, doi:10.22380/2539472X.795.

[49] A. Felicien, 'Resistencias y disputas alrededor de las semillas en Venezuela', in *Semillas y crisis alimentaria en Venezuela*. Quito: Fundación Rosa Luxemburg Oficina Región Andina, 2021.

[50] P. Singh, 'Politics of knowledge in development: Explorations in seed sovereignty', *Studies in Indian Politics*, vol. 9, no. 1, pp. 105–117, Jun.2021, doi:10.1177/2321023021999179.

[51] D. Silva Garzón and L. Gutiérrez Escobar, 'Revolturas: Resisting multinational seed corporations and legal seed regimes through seed-saving practices and activism in Colombia', *Journal of Peasant Studies*, vol. 47, no. 4, pp. 674–699, 2020, doi:10.1080/03066150.2019.1668780.

2 Transmodernity and *Buen Vivir*

Transmodern(ity) is an umbrella term encompassing similar, emerging intellectual projects around the world, including the work of Barfield, Dussel, Eisler, Grossberg, Luyckx Ghisi, Rodríguez Magda, Venn, Wilber, and others [1], [2]. Rodríguez Magda (who appears to be the first to use the term in the literature) has also compared her work to Bauman's Liquid Modernity, Beck's Second Modernity, Jameson's Late Capitalism, Lipovetsky's Hypermodernity, and Žižek's Desert of the Real [3].

Ateljevic finds commonalities in most of these proposals, as they 'point to the same intuitive aspirations for inclusivity, diversity, partnership, sacredness and quality of life, deep play, sustainability, universal human rights, the rights of nature, and peace on Earth' [1, p. 46]. Some defining contours of Transmodernity are that it:

- Emerges from the realisation of the interconnected nature of planetary existence.
- Emerges from globalisation and information technology.
- Is the starting point of a new era of human consciousness, a new shared story of humanity.
- Is inherently decolonial, as it contests Eurocentricity.
- Contests Eurocentric truth claims but seeks to keep the best of Modernity.
- Is an integration/synthesis of the pre-Modern, Modern, and post-Modern.
- Is post-secular, re-engaging with spirituality but rejecting absolutist dogma.
- Is dialogically universal.
- Moves from geopolitics to biopolitics.

For Rodríguez Magda, Transmodernity is 'firstly, a description of a globalised, rhizomatic and technological society, gestated from the First World faced to its others, penetrating and assuming them. Secondly, it is the effort to transcend this enveloping, hyperreal and relativist closure' [3, p. 3]. In their work, Transmodernity is not a dialogue of civilisations, but a totalising chaotic whole in the new grand narrative (grand fact) of globalisation. In its multiple dimensions, this reality is characterised by transnational

DOI: 10.4324/9781003715092-4

organisations, problems, events, communities, and structures. Globalisation substitutes territoriality for cyberspace, where the local and the global coexist as the glocal; Transmodernity thus 'imposes a transcultural trans-ethnicity that is yet to build its own cosmopolitics' [3, p. 10].

Rodríguez Magda excludes non-Modern worldviews in her work, framing Transmodernity as a synthesis between Modernity and Post-Modernity. However, there are other conceptions where Transmodernity also affirms 'the essential components of modernity's excluded cultures in order to develop a new civilisation for the twenty-first century' [4, p. 224]. That is, the peripheral Other reifying their existence, enabling the re-emergence of deprecated cultures, and establishing critical dialogues between hegemony and alterity [5]. For Dussel, the hegemony of Europe has only lasted two centuries, and in that time it has been unable to transform ancient cultures radically [6]. These surviving cultures are simultaneously Pre-Modern, Modern, and post-Modern, despite countless colonising efforts over their ways of being and knowing.

Dussel defines Transmodernity as a future culture that assumes the best of Modernity to produce a pluriversal utopia through authentic intercultural dialogue. However, engaging in such symmetrical multiculturalism is extremely difficult and might even be impossible [6]. Borderline thinking (the space between one's own peripheral culture and Modernity) is proposed as the loci of critical engagement. For Dussel, Transmodernity is: (i) an affirmation of devalued cultural moments that are exterior to Modernity; (ii) an internal critique of peripheral cultures from traditional values, done by people living in the biculturality of the borders; and (iii) a long period of resistance and maturation, developing cultural traditions on the path towards a Transmodern utopia [6].

Dussel's work has been criticised for not escaping Modernity, as it insists on producing macro-identities and metanarratives [7]. However, Martín Alcoff states that Transmodernity emerges from a retelling of world history from the perspectives of the Others that Europe rendered invisible [7]. As a result, a perpetually decentred interconnected history arises. In this sense, Martín Alcoff interprets Transmodernity as allowing inclusivity of multiple modernities[1] without their totalisation. For Martín Alcoff, 'the transmodern meta-narrative suggests a recipe for moving forward ... via an analysis of how and where cultural dialogues can occur most productively given the way in which the current global discursive regimes have been affected by colonialism' [7, p. 65].

In a similar vein, Grosfoguel [8] sees Transmodernity as the possibility of transcending the theo-politics and consequent ego-politics of Eurocentric thought, a way of knowing that excluded and marginalised through epistemological racism. Transmodernity then can be a 'strategy or epistemic mechanism towards a decolonised Transmodern world that moves us beyond Eurocentric-First-World and Eurocentric-Third-World fundamentalisms' [9, p. 211].

Martín Alcoff, Rodríguez Magda, and Misoczky warn of the risk of fetishising the local and hyper-dimensioning community. Rodríguez Magda specifically criticises proposals for a return to an identitarian pre-Modernity [3]. However, it is important to understand that contemporary Indigenous and peasant movements (the non-Modern) are not struggling only for an affirmation of identities. They are engaging in a Transmodern dialogue that also wants the best of Modernity [5]. Rodríguez Magda might be falling for the old Western assumption that these peri-Modern sites are 'dead or epistemically barren, [when in actuality] they are alive with a distinct difference barely legible to the [Modern/Colonial core]' [7, p. 66].

For Cubillo-Guevara and Hidalgo-Capitán, there are four cosmovisions coexisting in the 21st century: pre-Modernity, Modernity, post-Modernity, and Transmodernity [10]. These cosmovisions coexist as competitive cultural paradigms, related by an evolutive alterity instead of a paradigmatic overcoming. The authors state that Transmodernity

> emerges as a reaction and negation to the evolutive alterity of Pre-modernity, Modernity and Postmodernity. It pursues a synthetical triangulation between these cosmovisions, assuming and integrating precepts from them ... The prefix 'trans-' denotes an intent to overcome these cosmovisions, going 'beyond' them while integrating them, taking elements from each by traversing them [10, p. 153].

For Cubillo-Guevara and Hidalgo-Capitán, Transmodernity is a

> worldview based on intersubjective consensus [looking] for consensual truths which can be apprehended by second-order research. This cosmovision ... interprets all aspects of life from the emotionally intelligent combination of postulates based on faith, reason and imagination. It pursues the realization of individual expectations through the participative construction of consensual projects that are socially and environmentally harmonic [10, p. 136].

In Transmodernity, communities (both territorialised and virtual) construct their own metanarratives of social transformation. Cubillo-Guevara and Hidalgo-Capitán propose Transdevelopment as a new wellness paradigm, triangulating pre-Modern, Modern, and post-Modern worldviews [10]. Its intention is to build an alternative that goes beyond subsistence, Development, and post-Development. It pursues the satisfaction of the material and immaterial needs of society, through participatory processes, where needs and means to satisfy them are decided under principles of social equity and environmental sustainability. This implies that each community can define their own understanding of wellness, keeping nature and society at the centre of individual concerns. In Transdevelopment, individuals feel part of different communities and of nature, which they might consider as a living entity, even attributing to it qualities of consciousness or divinity [10].

Transdevelopment has at least two converging efforts at social transformation, which can be considered as a complementary dyad: degrowth (Global North/urban) and *Buen Vivir* (Global South/rural). However, these are not exclusive, as other Transmodern efforts are being mobilised around the world, for instance *ubuntu* and *ukama* in Africa [11], [12].

The term degrowth emerged at the turn of the century[2] as a 'provocative slogan, to denounce the mystification of the ideology of sustainable development' [13, p. 1]. Degrowth focuses on the reduction of production, consumption, and accumulation to enhance wellness [10], since growth is seen as increasingly toxic and happening at the expense of nature, future generations, and the peoples of poor nations [14].

Degrowth can be defined as 'a socially sustainable and equitable reduction (and eventually stabilisation) of society's throughput'[3] [15, p. 874]. In their analysis of limits to growth literature, Beling and colleagues mobilise three arguments for degrowth: (i) the unsustainability of Western standards of affluence; (ii) the historical unfeasibility of achieving them; and (iii) the general unhappiness found in most affluent countries [16]. Degrowth thus aims to rupture stablished ways of being through a virtuous cycle of reassessing, reconceptualising, restructuring, relocating, redistributing, reducing, reusing, and recycling [14].

Degrowth seeks to transform the current system to promote balance between material and non-material prosperity through 'time prosperity, "relational goods" (friendship, neighbourliness, etc.), non-capitalistic, community-based forms of production, exchange, and consumption, [among others]' [16, p. 307]. Degrowth can be framed as a nature-inclusive socialism. It is a transformative political vision that is considered left-wing since it criticises consumer societies and liberalism while having utopian aspirations [15], [17]. However, it is different from left-wing traditions like the 'social-democratic, communist, Trotskyist, and others which have fallen into the trap of the Keynesian-Fordist compromise … seduced by the myth of the pie that grows indefinitely' [14, p. 55].

This exploration of degrowth is by no means exhaustive. Research on degrowth is vast, as it involves history, economics, anthropology and social sciences, ecological economics, political science, and technology studies [18]. As this research focuses on transformations in the Andean region, it will expand on the Latin American transformative project of *Buen Vivir. Buen Vivir* can be considered Transmodern as it is inclusive and eco-centric, advocating for rights of nature, intercultural dialogue, and social and environmental harmony, while challenging Eurocentricity.

Buen Vivir

Buen Vivir (BV) is an emerging Latin American concept of wellness that can be translated as living well, living the good life, plentiful life, knowing how to live, or good living. It is still a concept under construction, with several interpretations [19], but in its most basic form, it refers to living in harmony with oneself, community, and nature [10], [20]. The concept is a contemporary interpretation of ancient Amazonian Indigenous understandings of wellness, crystallised in the Quichua neologism *Sumak Kawsay* [21]. The main proposals of living well are the Ecuadorian 'Buen Vivir' (from the Quichua *Sumak Kawsay*) and the Bolivian *Vivir Bien* (from the Aymara *Suma Qumaña*).

BV's focus on identity, equity, and sustainability can be expanded by defining some of its principles [22]:

- Coexisting in multipolarity: the whole is multipolar; this includes understanding how to live in communities of difference, in nature, and that inequalities will always exist.
- Dynamic equilibrium: despite not being egalitarian, there is a constant search for balance of the whole. Humans play the role of intermediaries in the quest for equilibrium by connecting with nature, but also trying to balance the spiritual with the material, knowledge and wisdom, different cultures, etc.
- Complementarity: recognition of 'the possibilities diversity provides for balancing and integrating the contradictions of the whole' [22, p. 4]; a constant effort to learn from difference, leading to the existence of *Buenos Vivires*,[4] or plural futures from diverse wisdom(s), knowledge(s), and practices.
- Decolonisation: a constant effort to dismantle the continuing domination of the Eurocentric worldview, leading to self-management and self-determination of communities and territories.

Adding to these attempts to define the concept, Beling and colleagues identify four constitutive elements of BV: (i) the idea of harmony with nature, including its abiotic components; (ii) vindication of the principles and values of marginalised/subordinated peoples; (iii) the state as guarantor of the satisfaction of basic needs, social justice, and equality; and (iv) democracy [16].

While this perspective gives a central role to states, Solón puts greater emphasis on communities, social movements and organisations [22]. These actors must play a preponderant role as state implementations of BV have until now been more reformist than transformative [23]. In fact, BV has always been developed from the base, leading to the empowerment of social movements to be more creative, resilient, and self-organised. In grassroots understandings of BV, direct democracies are preferred, and the state plays the role of a facilitator. Its mission is the socially conscious deployment of technological advancements while coordinating local networks of production, exchange, credit, traditional knowledge, and innovation [22].

These diverse interpretations of BV have led some authors to map the field, seeking commonalities, and identifying different strands of BV in Latin America [19], [21]. For Chassagne, they are: extractivist, where governments use extractivism to fund/achieve BV; and ecological, informed by Indigenous communities and post-Development literature [19]. For Hidalgo-Capitán & Cubillo-Guevara [21], BV strands are:

- Indigenist and Pachamamist (from Pachamama, the Andean goddess of nature) – intellectuals in this camp want to recreate the harmonious past living conditions of Indigenous communities in present-day Latin America, prioritising Indigenous identities.
- Socialist and statist – intellectuals in this camp seek the implementation of 'Twenty-first century socialism … with modern development in its neo-Marxist form' [21, p. 28], prioritising equity to achieve some form of post-capitalism with the state as its main agent.
- Ecologist and post-developmentalist – intellectuals in this camp see communities as the main unit of change. Each territory and community defines its own concept of good living, with nature at the centre of their concerns. It prioritises sustainability as understood by deep ecology.

Transcending the many definitions of BV and their diverging intellectual influences, authors call for a synthesised concept of BV. This definition 'would contribute to transforming Latin American societies into pluri-national, post-capitalist, and biocentric societies under a trans-modern and transdevelopment conception that transcends … paradigms of well-being, such as pre-modern subsistence, modern … and postmodern development' [21, p. 26].

BV could then be considered as a 'practical tool or resource for endogenous, community-led change' [19]. Instead of an ideological alternative to dominant paradigms, BV could be a platform for plural alternative visions and ways of collaborating to develop effective solutions.

After more than 35 years from the emergence of the BV concept, it has effectively influenced national-level policy and other phenomena in Latin America (see Part 2 of this book). However, there are still challenges for its implementation [19]:

- The contested nature of the concept.
- The need for multiple perspectives to come together in practice (institutions and communities, macro and micro levels).
- Political attribution of the concept.
- The existence of BV in extractive economies.
- The need to develop indicators to inform debate, and support implementation and measurement.
- The need to understand the core principles of BV in practice.

The *Buen Vivir* concept, its evolution and transformative potential, will be explored further in subsequent chapters. For some, BV is so broad that it lacks any real meaning; for others, it is a utopia impossible to attain. Its nature-centric approach and inclusion of personhood or rights of nature have been condemned by both right- and left-wing politicians, framing it as an infantile strategy of indigenist politics [23], [24], [25]. However, for many more in Latin America (mainly historically marginalised peoples), *Buen Vivir* has opened new transformation pathways, producing real impacts at multiple levels.

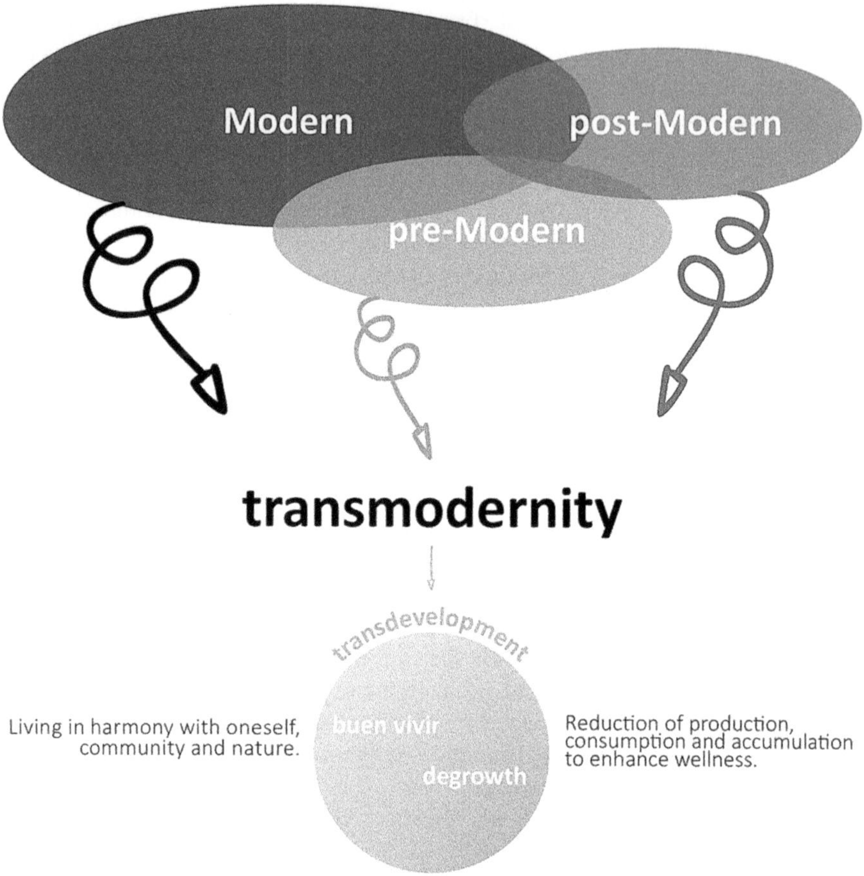

Figure 2.1 Transmodernity and the Transdevelopment dyad

Transcending the abyssal line

An essential part of the Transmodern and *Buen Vivir* intellectual projects is to provincialise the Eurocentric tradition to transcend the abyssal line. For Santos, this process begins by recognising our own blindness, so that we can move towards an epistemology of seeing [26]. This way of knowing calls for recognition and connection to the absent knowledge(s) and absent agents who were pushed to the peripheries by Modernity/Coloniality. It is also about understanding social practices as knowledge practices, so that non-scientific practices can be understood as alternative rival knowledge(s).

An epistemology of seeing is ultimately about enriching 'the parameters that define authorship, intelligibility and purposefulness … [while submitting] narrowly defined technological applications of knowledge to political and ethical contestations' [26, Ch. 4]. This transformation from blindness to seeing requires new relations between knowledge and politics. Ideally, this will produce a new epistemological stance that embraces a plurality of knowledge(s) and practices.

This transformation is not about a more equal distribution of scientific knowledge, but rather the production and sustainment of contextualised knowledge(s) that become trans-scalar through subaltern linkages at global scale, thus becoming glocal. It is not about discrediting scientific knowledge, but rather its mobilisation in a counterhegemonic manner. For Santos, the stance of epistemologies of the South is achieved through two procedures: ecologies of knowledge(s) and intercultural translation [26].

Ecologies of knowledge(s), knowledge constellations, or interknowledge aim to create new meaning articulations through sfumato, mestizaje, bricolage, or transculturation. These are integrative hybridising processes, where lines are blurred to enrich parameters of intelligibility and purposefulness. An ecology of knowledge(s) is premised on radical co-presence, which is the recognition of the contemporaneity of agents and practices on both sides of the abyssal line, requiring a Transmodern decolonial rejection of any conception of linear time.

Interknowledge can be considered a utopia of constant learning, where people engage with less familiar knowledge(s) without forgetting their own knowledge or older ones. It also mobilises knowledge-as-emancipation, where the purpose of knowing is solidarity, or the 'recognition of the other as an equal and as an equal producer of knowledge' [26, Ch. 5]. This solidarity has a caveat: people should be recognised as 'equal, whenever difference makes [them] inferior, and as different, whenever equality jeopardises [their] identity' [26, Ch. 5].

The second procedure of epistemologies of the South is intercultural translation; a 'living process of complex interactions among heterogeneous artifacts, both linguistic and non-linguistic, combined with exchanges that by far exceed logocentric or discourse-centric frameworks' [26, Ch. 8]. The practice of translation would allow for 'mutual intelligibility among culturally diverse social experiences of the world' [26, Ch. 8]. Translational contact zones are proposed as time-spaces of mediation and negotiation where relational inequality is reduced. These contact zones are

> social fields in which different cultural life worlds meet, mediate, negotiate and clash … In which rival normative ideas, knowledges, power forms, symbolic universes and agencies meet in usually unequal conditions and resist, reject, assimilate, imitate, translate and subvert each other, thus giving rise to hybrid cultural constellations in which the inequality of exchanges may be either reinforced or reduced [26, Ch. 8].

Translation work mobilises epistemological and democratic imagination with the objective of creating new constellations of knowledge(s) and practices able to challenge current ways of being and provide viable alternatives. Santos conceives translators as 'good subaltern cosmopolitan intellectuals' that are 'deeply embedded in the practices and knowledges they represent, having both a profound and critical understanding [of them]' [26, Ch. 8].

This book argues that translational contact zones have opened in three national seed systems (Colombia, Ecuador, and Venezuela). In these cases, social movements and state actors have served as translators that expand the parameters that define authorship, intelligibility, and purposefulness of seed-related knowledge(s) and practices. This has created an interesting and still ongoing dynamic, where new knowledge constellations may open transition pathways to alternative food futures that are more Transmodern, sustainable, and just.

Notes

1 'Multiple modernities' may be interpreted as different articulations of Pre-Modern, Modern, and Post-Modern worldviews.
2 See S!lence N°280 'La décroissance La décroissance Décroissance soutenable et conviviale'. Available: https://www.revuesilence.net/numeros/280-La-decroissance/?vue=telecharger.
3 The materials and energy extracted, processed, transported, and distributed to consume and eventually become waste.
4 *Buenos Vivires* is the plural form of the phrase 'living well' in Spanish.

References

[1] I. Ateljevic, 'Transmodernity: Integrating perspectives on societal evolution', *Futures*, vol. 47, pp. 38–48, 2013, doi:10.1016/j.futures.2013.01.002.
[2] M. Blaser, 'Political ontology: Cultural studies without "cultures"?' *Cultural Studies*, vol. 23, no. 5–6, pp. 873–896, 2009, doi:10.1080/09502380903208023.
[3] R. M. Rodríguez Magda, 'Transmodernidad: Un nuevo paradigma', *TRANSMODERNITY: Journal of Peripheral Cultural Production of the Luso-Hispanic World*, vol. 1, no. 1, pp. 1–13, 2011.
[4] E. Dussel, 'World-system and trans-modernity', *Nepantla: Views from the South*, vol. 3, no. 2, pp. 221–244, 2002.
[5] M. C. Misoczky, 'World visions in dispute in contemporary Latin America: Development x harmonic life', *Organization*, vol. 18, no. 3, pp. 345–363, 2011, doi:10.1177/1350508411398730.
[6] E. Dussel, 'Transmodernity and interculturality: An interpretation from the perspective of philosophy of liberation', *TRANSMODERNITY: Journal of Peripheral Cultural Production of the Luso-Hispanic World*, vol. 1, no. 3, 2012.
[7] L. Martín Alcoff, 'Enrique Dussel's Transmodernism', *TRANSMODERNITY: Journal of Peripheral Cultural Production of the Luso-Hispanic World*, vol. 1, no. 3, pp. 60–68, 2012.
[8] R. Grosfoguel, 'Hacia un pluriversalismo transmoderno decolonial', *Tabula Rasa*, no. 9, pp. 199–215, 2008, doi:10.25058/20112742.345.
[9] R. Grosfoguel, 'Transmodernity, border thinking, and global Coloniality – Decolonizing political economy and postcolonial studies', *Eurozine*, pp. 1–23, 2008. [Online]. Available: www.eurozine.com/transmodernity-border-thinking-and-global-coloniality.

[10] A. P. Cubillo-Guevara and A. L. Hidalgo-Capitán, 'El trans-desarrollo como manifestación de la trans-modernidad. Más allá de la subsistencia, el desarrollo y el post-desarrollo', *Revista de Economía Mundial*, vol. 41, pp. 127–158, 2015.

[11] M. Swilling, *The age of sustainability: Just transitions in a complex world.* New York: Routledge, 2020.

[12] S. H. Kumalo, 'Problematising development in sustainability: Epistemic justice through an African ethic', *Southern African Journal of Environmental Education*, vol. 33, no. 1, p. 14, 2017, doi:10.4314/sajee.v33i1.2.

[13] S. Latouche, 'Une société de décroissance est-elle souhaitable?' *Revue Juridique de l'Environnement*, vol. 40, no. 2, pp. 208–210, 2015, doi:10.3406/rjenv.2015.6697.

[14] S. Latouche, 'La décroissance est-elle la solution de la crise?' *Ecologie & politique*, vol. 40, no. 2, p. 51, 2010, doi:10.3917/ecopo.040.0051.

[15] G. Kallis, 'In defence of degrowth', *Ecological Economics*, vol. 70, no. 5, pp. 873–880, Mar.2011, doi:10.1016/j.ecolecon.2010.12.007.

[16] A. E. Beling, J. Vanhulst, F. Demaria, V. Rabi, A. E. Carballo, and J. Pelenc, 'Discursive synergies for a "great transformation" towards sustainability: Pragmatic contributions to a necessary dialogue between human development, degrowth, and *Buen Vivir*', *Ecological Economics*, vol. 144, March, pp. 304–313, 2018, doi:10.1016/j.ecolecon.2017.08.025.

[17] S. Latouche, 'La décroissance comme projet politique de gauche', *Revue du MAUSS*, vol. 34, no. 2, pp. 38–45, Oct.2009, doi:10.3917/rdm.034.0038.

[18] G. Kallis, V. Kostakis, S. Lange, B. Muraca, S. Paulson, and M. Schmelzer, 'Research on degrowth', *Annual Review of Environment and Resources*, vol. 43, pp. 291–316, 2018, doi:10.1146/annurev-environ-102017-025941.

[19] N. Chassagne, 'Sustaining the "good life": *Buen Vivir* as an alternative to sustainable development', *Community Dev J*, vol. 54, no. 3, pp. 482–500, 2019, doi:10.1093/cdj/bsx062.

[20] E. Gudynas, '*Buen Vivir*: Today's tomorrow,' *Development*, vol. 54, no. 4, pp. 441–447, 2011, doi:10.1057/dev.2011.86.

[21] A. L. Hidalgo-Capitán and A. P. Cubillo-Guevara, 'Deconstruction and genealogy of Latin American good living (*Buen Vivir*). The (Triune) good living and its diverse intellectual wellsprings', in *Alternative Pathways to Sustainable Development: Lessons from Latin America*, G. Carbonnier, H. Campodónico, and S. T. Campodónico, Eds. Leiden: Brill Nijhoff, 2017, doi:10.1163/9789004351677.

[22] P. Solón, '*Vivir Bien*: Old cosmovisions and new paradigms', February2018. Available: https://greattransition.org/publication/vivir-bien.

[23] K. Weld, 'Holy war: Latin America's far right', *Dissent*, 2020. [Online]. Available: https://www.dissentmagazine.org/issue/spring-2020.

[24] M. de la Cadena, 'Indigenous cosmopolitics in the Andes: Conceptual reflections beyond "politics"', *Cultural Anthropology*, vol. 25, no. 2, pp. 334–370, 2010, doi:10.1111/j.1548-1360.2010.01061.x.

[25] M. Tola, 'Between Pachamama and mother earth: Gender, political ontology and the rights of nature in contemporary Bolivia', *Fem Rev*, vol. 118, no. 1, pp. 25–40, 2018, doi:10.1057/s41305-018-0100-4.

[26] B. de S. Santos, *Epistemologies of the South – Justice against epistemicide.* New York: Routledge, 2016.

Part 2

Andean seed system change

3 Andean seed guardian networks

> I don't know if she told you the story of when she went to an [agroecology] event in Manizales. On the way back someone got off before [her] … and took her suitcase. Oh! What a tragedy that was, she called us extremely upset, almost crying. But it wasn't because she had clothes or money [in the suitcase], but because she had seeds that she had exchanged with Indigenous peoples. That's the level of appreciation [she has for seeds], thank goodness she was able to recover it [1].
>
> [Venezuelan Indigenous Guajiro peoples] are the ones who have really done the conservation [of porcelain cacao] … through thick and thin. Because we're talking about a dismantled jungle matrix, completely converted to cow pasture. And their orchards are little islands, groves in the middle of those cattle ranches. The passion, the authenticity with which [a Guajiro elder] spoke to me about that cacao and his work to maintain it, and his concern about what was going to happen to the grove when the head of the matriarchal line died … And he also talked about the continuous visits he had from Japanese and American gentlemen. 'These *gringos*[1] … they came to ask me to do them a little favour and give them some of those [cacao] shells. And you know I'm not selfish, I gave it to them, and they brought me a few things, they gave me a hat'. In other words, he did it in reciprocity [2].

These stories succinctly express the spirit of seed guardians,[2] particular type of knowledge-holder who specialises in caring for one or several seed varieties that are creole, native, or landrace. The above extracts paint an initial picture of a group of people so widely diverse that it complicates definitions. Due to their diversity, this chapter will focus not on developing a single definition of seed guardians, but on mapping their shared narratives and practices. Data is drawn from interviews with eight members of networks operating in Colombia, Ecuador, Mexico, and Venezuela, as well as document analysis from publications across different media. Pseudonyms are used throughout the chapter to anonymise interviewees. Although differences across seed guardian networks and the nations in which they operate are vast, their commonalities are emphasised.

DOI: 10.4324/9781003715092-6

As expressed by Pedro when asked about Ecuadorian seed guardians, they

> are men, women, Americans, Canadians, French, Belgians, Peruvians, Colombians who live here, and obviously mostly Ecuadorians. But among Ecuadorians there are Afro-descendants, peoples of various Indigenous nations, coastal people, mountain people, from the Amazon. Even ideologically, … we have people who are very Catholic, we have evangelicals, atheists, communists, we have people from the right and the left. It's a mishmash … we have a colleague who is a shaman and several colleagues who are scientists [3].

When asking a peasant farmer, they will say that guardians are 'peasants but at the same time involved with biodiversity; [seeking] that ancestral and native seeds are not lost [in a food sector where] there are seeds that no longer reproduce'[3] [4]. More urban-based and professional guardians speak of themselves as defectors from the workforce, abandoning the rat race to pursue a 'post-industrial mode of life: A return to craft, autonomy, and the strengthening of communities' [3].

In truth, seed guardians come in all shapes and sizes, they live in diverse landscapes, and span many biological phenotypes, ethnicities, age groups, genders, and classes. Indeed, the only widespread commonality might be a 'desire to provide a positive alternative to the current system in harmony with nature and seeking social justice' [3].

This pursuit of alternatives makes guardians feel like outliers, not only in the Modern/Colonial system but also amongst their more non-Modern peers. As Carmen expresses it:

> at first it was difficult for [guardians] to even admit that they had those seeds. Because keeping them was tied to attitudes not in line with progress, modernisation, [and] monetisation. It was something more akin to poor, stubborn, and ignorant peasants who have useless seeds that nobody is interested in … and by nobody, I mean the market [2].

Beyond this condemnation from Modern/Colonial perspectives, guardians are also extremely unusual in non-Modern spaces; so unusual that they need to incentivise conversations with their peasant neighbours who do not understand why they hold so many seeds. As Pedro explains, there are 'several things about seed management that only guardians understand', adding that the networks often 'have to pursue [other] peasant and Indigenous [social] movements to explain seeds because they don't understand them' [3].

Feelings of distinctiveness at the local level are reflected in testimonies from Ecuador like those where '[guardian] brothers and sisters have suffered attacks by neighbours for walking this alternative path, [with] many feeling deeply discouraged by being so isolated in their territories' [3]. This sense of dislocation is further explained by mestizo neo-peasant[4] Pedro in relation to his friendship with an Indigenous seed guardian:

> both of us are black sheep in our original social environments. My wife and I, for many things, but one of the main ones is our love for the Indigenous and ancestral. While my *compadre*[5] and his wife, … they're black sheep within their Indigenous context because they quote unquote have westernised a lot [3].

This sentiment is shared by Colombian guardian Yolanda, who sometimes feels uneasy engaging in guardianship practices, stating that: 'when [neighbours] come to my house for something, … sometimes they look at me like I'm weird, [like they're asking themselves] why does she grow all that?' [4]. Indeed, Yolanda is a special kind of guardian, caring for more than 50 plant varieties (proudly mentioning them one by one during the interview). Such an amount is clearly above average, giving a sense that this guardian is more akin to an obsessive collector – an uncommon behaviour, as most guardians focus only on a specific variety or a small number of viable plants in their territory.

Despite her clear particularities, Yolanda links the uneasiness of her neighbours with an increasingly common attitude amongst farmers: 'what isn't sellable shouldn't be produced' [4]. Beyond this peripheral position to the industrialist system, Yolanda also gives a sense that seed guardians are differently related to non-humans in their territories. She likes putting photos of her plants on social media, refers to bees under her care as 'babes', and is thankful for the help of birds in disseminating corn seeds on her farm. Yolanda admits that she gets 'really sad' when she loses seeds, and when prompted to pick a favourite she is unable to do so, since 'they're all really important to [her]' [4]. These recognitions of agency, interrelatedness, and horizontal relations in nature further position guardians as non-Modern actors peripheral in the Modern/Colonial world system.

These alternative connections to nature are embodied by Indigenous, Afro-Latino, and peasant actors, but are also present in more urban and professionalised seed guardians. For instance, mestiza neo-peasant guardian Lina positions herself as agro-descendant, affirming that 'my grandparents, my family were peasants … and my grandparents were my first teachers' [5].

This was also highlighted by Pedro when he stated that 'like most Ecuadorians, [he] had grandparents who were peasants. In truth, almost all [Ecuadorians] are first, second, and maximum third generation peasant descendants' [3]. Memories and family histories tied to land and food production inform much of the activism of seed guardians. They actively connect seeds to politics, framing their own guardianship practice as political, and recognising the marginalisation of AfroIndopeasant identities throughout Latin American history. To some guardians like Pedro, changing these unjust structures is assumed as a responsibility.

The complexities of contemporary seed systems and the immense burden of transforming centuries-old structural injustices led these disparate actors to organise. Seed guardian networks (SGNs) emerged at the beginning of the 21st century in Latin America, responding to the intensification of neoliberal policies during the 1990s and early 2000s. These policies opened more spaces for transnational corporations while curtailing possibilities for local traditional agricultural practices (specially as part of free trade agreements with rich nations). This neoliberal surge also coincided with: (i) an increased awareness of the pitfalls of the green revolution; (ii) a positioning of agroecology and organic agriculture as viable responses; and (iii) the global spread of resistances to increasing regulation, privatisation, and monopolisation of seeds (e.g., seed freedom in India; see section below).

The mission of these networks can be broadly described as the defence of people's rights to produce, store, improve, exchange, and trade their own seeds according to Indigenous, Afro-descendant, peasant, and agroecological practices. Initially, the activities of SGNs were mostly directed towards resisting detrimental legislative changes and getting their voices heard. With time, those resistances allowed the flourishing of protected spaces for non-Modern seed practices. After nearly 20 years of work, the networks have consolidated as an influential political force opening spaces for the non-Modern by voicing the concerns of historically marginalised actors in policymaking. This organisational process has enabled the emergence of an influential trans-local innovative space that increasingly participates in seed system transformation.

SGNs are decentralised, place-based, trans-local coalitions of people and organisations committed to safeguarding care practices of sowing, reproduction, and exchange of landrace, native, and creole seeds. Participants are widely diverse but include peasant, Indigenous, and Afro-Latino communities, neo-peasants, foreign and national NGOs, urban academics and professionals, state bureaucrats, amongst others. Over the last two decades, and despite being under constant pressure from both government and industry, SGNs have connected otherwise isolated outliers to simultaneously resist Modern/Colonial food narratives, practices, and rules, while proposing alternatives.

The purpose of this chapter is to offer a characterisation of Andean seed sovereignty social movements, one of the main drivers behind the transformation processes described in this book. The chapter continues with a description of the global context in which these actors emerged, the narratives that provide meaning to their work, and their internal practices and relations to external actors.

Alternatives to industrialist food production

Andean SGNs are but a recent iteration of centuries of resistance to Coloniality. As explored in Part 1 of this book, the current food regime is facing a multidimensional crisis, and alternatives are emerging to transform it. Food systems in postcolonial nations can generally be interpreted as hybrids between Modern and non-Modern worldviews. Even with the massive efforts of the green revolution to industrialise global food systems, they remain Transmodern. However, there are deep power asymmetries that position the industrialist and capitalist food system as dominant despite large contributions of the informal(ised) system to global food security [6]. The informal(ised) food system keeps being ignored or made invisible by incumbents but has been empowered over the last decades. This section provides a summary of the alternative narratives, approaches, and social mobilisations that have consolidated as food system crises become more evident. These global and interconnected alternatives set up the stage for native seed saving networks to emerge and influence the broader food system.

Myriad alternatives

The dominant food system is in crisis. Despite increasing food production while keeping prices stable, it has failed to achieve food security while producing negative social and environmental impacts [7], [8], [9]. Accordingly, alternatives are emerging as urgent calls to transform the system are increasing [7], [10], [11]. There are many narratives, future visions, practices, and goals in this space, as well as coalitions working towards food system transformation globally. Prominent approaches to sustainable agriculture include agroecology, nature-inclusive agriculture, permaculture, biodynamic agriculture, organic farming, conservation agriculture, regenerative agriculture, carbon farming, climate-smart agriculture, high nature value farming, low external input agriculture, circular agriculture, ecological intensification, or sustainable intensification [12].

Defining each of these concepts exceeds the purpose of this book.[6] They are interconnected, often overlap, and definitions, relations, and hierarchies between them are still contentious. There are profound differences regarding the direction of change, and tensions between reformist and more radical approaches often impede coalition building [13]. In fact, most of these narratives have emerged from and been dominated by Global North and corporate actors tending to maintain the status quo [8].

This section will focus on regenerative agriculture (RA) and agroecology as they move beyond reformist approaches and include social concerns pertinent to Global South actors. RA and agroecology share some common elements, such as

> emphasizing ecological concepts and principles to food production, and farming with minimal degradation of soil, water, and ecosystem services while producing sufficient healthy diverse foods for consumption and livelihoods. Both overlap in the list of production practices that distinguish them from more industrial or conventional production [14, p. 3].

Although powerful actors tend to co-opt these narratives to maintain the current industrialist paradigm [14], [15], these concepts (particularly agroecology) have informed alternative food system transformations over the last decades.

Regenerative agriculture

There have been calls to transition to more regenerative food systems [7], [9], [11]. At its most basic, these systems can be defined as 'maintaining positive reinforcing cycles of wellbeing both within and beyond themselves, especially between humans and wider nature' [11, p. 2]. In a more expanded manner, regenerative food systems

> provide food for human use but also sequester carbon, preserve biodiversity, produce diverse diets to combat malnutrition, and build community well-being by maintaining farming livelihoods and the social reproduction of culture and farming communities, support the dignity and autonomy of the person and allow severing ties of dependence on external inputs and external knowledge [9, p. 18].

To produce these regenerative food systems, proponents call for a regenerative agriculture (RA). This concept emerged from the Global North but aims to transcend the bounds of industrialist food production. Although there is no set definition of RA and the term is applied liberally [8], [14], it is generally understood as practices or principles that go beyond sustainability, including some social elements related to the wellbeing of farming communities [8]. Regeneration here should be understood as maintaining healthy ecosystems by preserving interconnected ecological processes. In this sense it is not only about practices or goals but relates to an ethic of reciprocity [14].

Despite still struggling to recognise some inequities in food systems, RA has risen to prominence over the last decade [8]. It is important to clarify that this concept is sometimes used to group many other sustainable agriculture alternatives. However, this book avoids using it as an umbrella term, rather connecting it to agroecology and food sovereignty as evidence of a broader paradigmatic shift towards alternative food systems.

Agroecology

Ecological principles and practices have been used by Indigenous cultures in food systems for millennia. However, these practices were kept in the margins until the 1960s when a combination of the disciplines of ecology and agronomy mainstreamed them in science. Agroecology has become increasingly prominent over the last three decades as Indigenous and peasant movements gain more visibility in their resistance against the dominant food regime.

Agroecology seeks to apply ecological principles to localised food systems prioritising knowledge and agency of Indigenous and peasant actors. It seeks to move agri-food practices from an exploitative logic to a logic of care and maintenance [13]. This shift includes approaches like permaculture, organic or biodynamic farming, and stablishing direct links between producers and consumers, among other strategies [13], [16]. Although the movement is growing, these approaches are still marginal and lack funding from state and private aid actors [16]. However, agroecology has been somewhat institutionalised internationally due to social movement pressures [17].

Agroecology is not only concerned with growing food sustainably; it is also politically oriented and has more radical goals of social transformation (e.g., reducing inequalities) [14]. As such, it is said that agroecology is simultaneously science, practice, and social movement [14], [17]. Some authors claim that it is a political mega-movement as it groups large numbers of self-organised processes at global scale [17]. Many international and transnational coalitions promote agroecology. However, the most prominent and powerful is undoubtedly La Via Campesina[7] (the peasant way), which voices Indigenous and peasant concerns and has more than 200 million members in 81 countries [7], [17].

The agroecological movement rejects neoliberalism, and advocates for agrarian and land reform, and for food sovereignty (see section below). Its aim is to transform food production according to agroecological principles and to implement it via empowered peasant subjects [17]. Accordingly, its project could be framed as a 'repeasantisation' of agri-food systems [13]. Additionally, it seeks to reduce energy consumption and transform value relations by configuring solidarity economies [17]. Participatory processes and knowledge co-creation are also important in agroecology [14], particularly the creation and dissemination of knowledge amongst peasant peers through dialogues of knowledge(s) (*diálogo de saberes*) and the peasant-to-peasant method (*campesino-a-campesino*) [13].

These processes of knowledge-sharing and broad participation allow the movement to connect a wide range of actors and create internal cohesion. It has also shifted established understandings of food systems by states and development organisations. This has imbued the agroecological social movement with political power [17]. Transnational examples of their influence may include the Declaration on the Rights of Indigenous Peoples 2007 (UNDRIP) and the Declaration on the Rights of Peasants and Others Working on Rural Areas 2018 (UNDROP) [15], [16]. These international normative instruments were highly influential in the change processes described in subsequent chapters, particularly UNDROP, as it aims to expand peasant participation in decision-making and governance, and affirm peasants' right to seeds, land, and natural resources [15].

Food sovereignty

The concept of food sovereignty emerged in 1996 as a peasant-led response to persistent inequalities, neoliberal policies expanding corporate control over food production, and to the Declaration on World Food Security by the United Nations [13], [15], [18]. While food security focuses on access to sufficient, safe, and nutritious food, food sovereignty emphasises power relations, aiming for more distributed control and determination over the system [19].

The 2007 Nyéléni Declaration broadened the definition, pushing food sovereignty movements to promote deep transformations focused on the right of people to provide for themselves, including access (and redistribution) of land, seeds, and other inputs, possibly constituting an alternative development paradigm [15].

The movement is perceived as radical as it mobilises post-capitalist framings to implement alternative economic models [15], [18]. It pushes states to ensure access to 'safe, nutritious, and culturally appropriate food in a non-discriminatory, participatory, and accountable manner' [15, p. 172]. It seeks to eliminate oppression and inequality intersectionally [13]. It promotes regenerative agriculture and agroecological practices in food production, emphasising sustainability [15], and works in a participatory manner, co-creating through *diálogo de saberes* [18].

Diverse social movements have formed and coalesced around this concept. However, its main proponent is La Via Campesina (LVC), which affirms that state reforms maintain power imbalances and only a strong social movement can transform food systems [15]. LVC has put forward six pillars of food sovereignty: 'focusing on food for people, valuing food providers, localizing food systems, putting control locally, building knowledge and skills, and working with nature' [18, p. 72]. These and other food sovereignty agendas have been questioned regarding issues of gender inequalities, the dismissal of existing technologies, and difficulties engaging with markets at local and global levels [13].

The concept of food security has dominated governance and policy circles for decades. However, food sovereignty has been positioned as a political alternative by social movements grouping marginalised food system actors. These movements question the assumption that aid, technology, or increased food production will solve hunger. They not only resist neoliberalism and the commodification of food but produce contextualised change towards commoning, sustainable production, solidarity economies, and more transformative paths. As part of their agendas and the implementation of alternatives, participants of these movements also push for seed sovereignty as a critical component of food sovereignty.

Seed sovereignty

Historically, seeds were considered as common heritage of humanity but have been increasingly framed as privatised commodities since the 20th century [20]. This process has been met with resistance by those most affected, mainly peasant and Indigenous farmers.[8] As explained in Chapter 1, native seeds and peasant food production have been neglected by incumbents despite their importance for global food provision. What is referred to in this book as the informal(ised) food system may be responsible for most of global agro-biodiversity. By some estimates, informal(ised) actors care for 'seven thousand species of cultivated plants, including two million varieties, mainly local breeds, … 60,000 species of wild plants … 34,000 species of livestock and about 9,000 creole breeds' [17, p. 62].

For more than a decade, there has been ample evidence on the importance of farmer seed networks for agriculture [21]. However, native seeds and associated traditional sharing practices are under increased scrutiny and control by corporations and states. In this scenario of increasing neoliberal and ecological pressures over food systems, and with seeds being perhaps the most critical input, their protection has become more urgent. Accordingly, diverse overlapping networks and coalitions have formed to resist enclosure and defend rights to seeds [22].

The most prominent organisations working in this space are Navdanya[9] and La Via Campesina, founded between 1987 and 1992 [23]. To the best of the author's knowledge, there is no systematic review about localised seed saving initiatives at global scale. There is urgent need for synthesis of empirical cases across geographies. As these initiatives are widely diverse and localised, mapping them exceeds the purpose of this book. The focus will be on seed sovereignty movements operating in Latin America. To understand more about the global landscape of these movements, there are excellent researchers expanding knowledge in this space. For cases of Indigenous food and seed sovereignty in several countries see [19], [24], [25]. For seed activism in the Philippines see [26]. For links between seed sovereignty and agroecology in Colombia see [27]. For the role of libraries in seed lending in the United States see [28]. For knowledge politics in Indian seed sovereignty see [29].

The actions of such collectives are grouped under many names: seed defence, seed recovery, seed freedom, or seed activism (here referred to as seed sovereignty). More than a decade ago, Kloppenburg mapped a series of rights constitutive of seed sovereignty: (i) to save and replant them; (ii) to share them; (iii) to use them and breed new varieties; and (iv) to participate in related policymaking [23]. Seed sovereignty social movements have been involved in political and legal battles promoting collective rights over seeds and agrobiodiversity while establishing contentious relations with the state [22]. Participation of seed guardians in seed governance is of special interest for this book. Seed policy change processes with SGN involvement are detailed in Chapters 4–6. In broad terms, these processes entail a move from opposition to co-creative tension, leading to more Transmodern seed systems.

For seed sovereignty social movements, opposition has traditionally been directed towards intellectual property rights and genetically modified seeds. This is informed by understandings of genetic resources as a social product and common heritage that should be exchanged freely [23]. These postures have guided the affiliation of these movements to agroecology, the development of participatory breeding programmes, and their activism towards achieving legal sovereignty over seeds [23]. Commoning has also led the movement to create alternative markets, and other spaces beyond markets and the state that are usually self-governed [22].

How these postures and actions manifest in the Andean region will be further explored in this chapter. The processes of change towards legal seed sovereignty will be expanded in Chapters 4–6, where the capacity of these collectives to appeal to broad constituencies and their openness to other allies will become more evident. These emerging legal frameworks and micro-institutions allow for more traditional knowledge and native seed sharing, and facilitate collaboration between scientists and peasants to produce knowledge as proposed by Kloppenburg more than a decade ago [23].

Narratives of Andean seed guardians

The previous section presented the broad global context in which Andean seed guardians emerged and currently operate. The crisis of the current food regime has led to the emergence of alternatives like regenerative agriculture and agroecology. It has evidenced the need for food and seed sovereignty, which has manifested as localised initiatives that are trans-locally interconnected. The crisis and emerging alternatives of agroecology and food sovereignty provide a broad storyline that informs the actions and asserts the purpose of SGNs. This section emphasises the particularities of this storyline in the region as four interconnected narratives (see Table 3.1). These narratives have framed the rationale and scope of action, attracting diverse actors to produce change.

Table 3.1 Seed guardian narratives

Narrative	*Description*
Our seeds vs. their seeds	The tensions between understandings of seeds as living beings with agency or a commons, and seeds as privatised commodity subject to intellectual property rights.
Dependence vs. resilience	Seed guardianship is seen as an activity that promotes resilience. This relates to the crisis of the food system, the climate crisis, or weather variations that may affect crops. Corporate technological packages are seen as increasing dependence and disempowering farmers.
Recentring non-Modern institutions	As alternatives to the dominant food regime emerge, Indigenous and other traditional ways of producing food are revalued and mainstreamed. Although marginalised by incumbents' agro-industrial biases, guardians push for a plurality of approaches in food production.
Seed guardian futures	Moving towards regenerative food futures in which: (i) there is dignity for AfroIndopeasant, family, and small-scale producers, (ii) recognition of the importance of seed guardians and their networks, and (iii) broader transitions towards food sovereignty.

Our seeds vs. their seeds

Narratively, network participants understand seeds as the starting point of nourishment, but also as free, living entities that embody the origin and regeneration of existence. For them, humans and seeds have a deep relationship of nurturing and care, a connection that mobilises narratives of seeds as part of family and community. Participants understand seeds as an intergenerational commons beyond government and corporate control, sometimes going as far as speaking of seeds as entities with agency and rights of their own. They acknowledge the existence of plural nurturing practices, producing a wide diversity of agricultural seeds influenced by generations of localised knowledge(s), experiences, cultures, and traditions.

Guardians draw a clear line dividing native, creole, and landrace seeds from their industrial counterparts (especially those that are transgenic). Industrial or hybrid seeds are seen as a sterile commodity, which require agro-toxics to produce food and which die without reproducing. The spread of these technological assemblages produces a deterioration of the environment and increases concentration of power in the food sector. Guardians see this as a threat to their way of life and food security in general, a threat not acknowledged by food system institutions. This invisibilisation of risks is a product of deeply ingrained narratives of incumbent players that frame industrial seeds as good, while ignoring the potential (sometimes even existence) of AfroIndopeasant seeds as technologies.

Seeds in AfroIndopeasant hands

Guardians start by recognising seeds as a 'living being' [30], be it Yolanda saying that seeds are 'where life is encased' [4], Lina framing seeds as 'the origin of existence' [5], or Rita and Pedro claiming they are the 'beginning of' and 'regeneration of life itself' [3], [31]. They also mobilise narratives of seeds as having a spiritual dimension. Neo-peasant guardian Lina evokes Indigenous worldviews when she states that 'the seed represents the creation of the divine, the beginning of everything. That's where food comes from, where water starts, because [Indigenous peoples] talk about seeds in the plural, about human seed, food seed, and water seed' [5]. Similarly, for Carmen seeds are 'the blessing that no one is denied, because it is that wonderful life entity to which we owe physical care and spiritual reverence' [2]. For some, this spiritual connection between humans and seeds emerges because 'deep down we're all seeds' [3].

There is also a recognition of seeds as multidimensional, which is a shared perception in SGNs [32]. Indeed, when referring to peasant seeds, Rita sees them as a 'container of knowledge, experience, and culture, ... of generations, of a lot of work, of diversity' [31]. For Alma, seeds are also 'the origin of many processes [outside of] food production; processes of [personal and family histories]'. This guardian understands seeds as the origin, not only of food or life itself but of 'mobilisations, and struggles, connecting people and stories [to] ... walk around the world arguing [for transformation]' [32]. For Pedro, seeds also have an 'unparalleled metaphorical power', adding that 'when people talk about seeds ... in their heart, they're always referring to their own birth and that of their children, and to the hope that there's a future' [3].

The futural capacities of seeds are also echoed by Rita when she speaks of them as 'the basis not only of our food, but ... of our future' [31]. Emiliano elaborates this point further when affirming that 'without seeds it will be very difficult for humanity to continue existing' [30]. This guardian sees practices of seed adaptation as warranting food security for centuries to come. He believes that seeds are what really guarantees our planetary dwelling, and as such we should stop 'looking at nature over our shoulders, [recognising] that [seeds] are also living beings *allowing us* [emphasis added] to continue to exist on this planet' [30].

One of the most common narrative associations with seeds is freedom. Guardians speak of 'free seeds, seeds that circulate, a seed that nobody can appropriate' [2]. Indeed, a common motto of Latin American seed sovereignty social movements is that 'seeds should be able to walk freely' [4]. Ownership in that space becomes a complex issue, a struggle between nature, peoples, governments, and businesses. For some guardians like Pedro, seeds are the 'patrimony of peoples', and as such they 'cannot be a strategic resource of the state because the state did not create them, nor does it maintain them' [3]. At the same time, they should not be the property of a small number of private companies, or be subjected to intellectual property rights, or scientific intervention to modify their genes.

Of course, seeds are also linked to food and agriculture. Pedro claims that 'physically they represent the possibility of continuing our crops' [3], while Emiliano is keen on reminding us that 'agriculture is born from the relationship that communities have [built] while managing seeds' [30]. It is important to emphasise that guardians distinguish their agricultural practices from industrial production. Guardians seek to do agriculture 'from [their] own seeds, not from [those] that companies are creating to strengthen the green revolution model' [30].

This is a constant struggle, since incumbent actors in food systems frequently associate native, creole, and landrace seeds with low quality. Guardians are conscious that native, creole, and landrace seeds may carry disease or have lower yields than industrial varieties. This translates into an interest in developing and assuring good quality seeds, but 'not on the same terms that [governments] use with companies' [30]. Indeed, this tension over definitions of quality will inform much of the transformation processes analysed, especially as it relates to drawing a line between production and productivism.

Industrial seeds

Industrial seeds are frequently framed in opposition to native, creole, or landrace seeds in AfroIndopeasant hands. Some guardians go as far as to say that 'all those things that the native seed has, do not exist in industrial and transgenic seeds' [31]. Most concerns about industrial seeds come from the infamous 'terminator gene', which stops seed reproduction after one sowing cycle, a technological development that is nearly incomprehensible for guardians.

As expressed by peasant Yolanda, 'those [seeds] that come in packets, that are already certified … they get to a point where they die and never reproduce. I don't know why that is' [4]. Urban and highly educated guardians like Emiliano recall how companies call seeds 'merchandise, materials' [30]. Furthermore, Rita explains that these seeds are 'homogeneous commodities, appropriated and fetishised by capital … concentrated in a few hands of very large companies' [31].

Industrial seeds are not seen as singular technologies, but as an assemblage often referred to as a 'technological package', in which agrichemicals are indivisible from the seeds that require them to grow. Some guardians coming from an agroecological perspective challenge the use of words like 'chemicals', because they see it as a veiled way to refer to poison [30]. For guardians like Emiliano, these technological packages are 'degrading ecosystems, seeds, and people's health' due to the pursuit of profit by seed and agrichemical companies [30]. This leads to an increasingly accentuating global crisis of the green revolution model, which has not been fully recognised by big corporations [3], [30].

Industrial seeds and their continued use in tandem with agrichemicals are seen by guardians as a threat. For Rita, an industrialised agri-food system where power is highly concentrated is an 'example of what we do not want, it is a threat against diversity, health, [and] our culture' [31]. Peasant guardian Yolanda aligns with this position when she says that industrial agriculture 'isn't wellbeing, that's not having freedom as peasants, because we're stuck … if at some point [companies] tell us they won't sell to us anymore … we're stuck not eating more of that [food]' [4].

Dependence on hybrid seeds and agrichemicals is also seen as a threat by Pedro, but he also ties this to the loss of 'people adapting [seeds] to [their territory] so that [they] can have them … [Something] done by millions of peasants, for ten thousand years, in thousands of locations around the world' [3].

Threats also come in the form of the technological intervention of seeds themselves. For Emiliano, 'there is a sector of humanity that thinks everything can be synthetic, that everything can be created … and manipulated … and transformed. [I think we] must be very careful … because seeds are the very essence of life' [30]. Adding to this, Rita raises the stakes even further when she affirms that 'we're talking about companies having power over life … Basically, you could say that [this power] is the beginning of death' [31].

Finally, there is an awareness of asymmetrical narratives in dominant food system institutions differentiating industrial and AfroIndopeasant seeds. For Emiliano, 'institutional frameworks ignore native and creole seeds, and instead of promoting and valuing them, they degrade them and try to keep them from circulating because they say they're of poor quality, that they're not productive' [30]. This negative image of native and creole seeds is of course the opposite of what guardians seek. However, in a space dominated by agro-industrial visions, it becomes almost impossible to circumvent established narratives.

In fact, Carmen affirms that seeds from transnational corporations circulate under a 'discourse that it is *the* [emphasis added] seed, the only one there is' [2]. For guardians, narratives of uniqueness and superiority of hybrid seeds in the food system are unfounded. Emiliano explains it in relation to racism:

> It's like saying black is bad because it's black and white is good because it's white, that's not what it's about. That's the problem with this whole issue of seeds, they simply say that [seeds] from the companies are good, and native and creole seeds are bad, because [the first] are produced by whites and the others are produced by blacks [30].

Dependence vs. resilience

As previously mentioned, guardians are deeply concerned with corporate control over food systems. Peasant guardian Yolanda sees control manifesting at multiple levels: people's dependence on supermarkets, national dependence on hybrid seed imports, and seed regulation (especially certification schemes and patents) curtailing spaces for peasant seeds. For her,

> people have been patenting [our seeds], taking them and putting ownership over them, [they] say, these are mine, no one can have them but me, only I can sell them. And seeds must be free, they must circulate freely amongst peoples, among communities [4].

Rita shares this sentiment when she states that seed defence is essential, since industrial certified seeds are a 'threat against us, our life and culture for the future' [31]. Seed regulation is always seen as pitted against Indigenous, Afro-Latino, and peasant peoples, and almost always heavily biased towards agro-industry. Alma visualises this as governments that usually favour elites, leading to the creation of 'anti-peasant laws' that criminalise traditional practices [32]. Or, as peasant guardian Yolanda puts it, 'what is ours, what is traditional on the land, is supposedly illegal to do. What I do, in some way, it's not right according to the law' [4].

For Yolanda, seed guardianship is important 'to not eradicate our seeds; ours, as well as those of the country, [to have] something to leave for future generations' [4]. Somewhat aligned with this futural position, seed guardianship is also seen as a defence of biocultural diversity against increasing homogenisation of food products. For Rita,

> if the defence and the fight for agro-biodiversity does not continue in peasant hands, we are going to lose that diversity in the future, [and that] would totally change our diet. I sometimes see agro-industrial production … like trying to put nature in a box. And diversity you cannot put in

> a box, if you put it in a box it will die, it cannot breathe, it cannot continue evolving, so … the defence of seeds is essential [31].

This is echoed by Yolanda when she speaks of the importance of joining the seed network:

> it is to save the biodiversity that we have as patrimony of the world, because with so many laws, and changes that have been made to seeds themselves that they don't reproduce … That made me think that we cannot always depend on multinationals to maintain those seeds. You don't know what they give you, they just don't work to continue sowing again, but if you keep your native and creole seeds, those seeds do [4].

Some guardians magnify this line of thought to apocalyptic levels, particularly in Ecuador where guardians feel that food systems are experiencing a major collapse. This is evident when Pedro expresses that 'we are now living in a moment in which we realise that this world is no longer possible … This can't go on, we must produce change' [3]. Acknowledgements of collapse do not fully translate to despair, as yet another common narrative frames seed guardianship as a hedge against climate chaos. This is deeply tied to narratives of free circulation and the practice of seed exchange amongst guardians.

As Lina puts it,

> today I planted this seed that I got from a guardian, and tomorrow if that guardian for some reason loses their seed because of a frost, [or something else], it turns out that sharing that seed … with me, … guarantees they can get it back [5].

This narrative is reproduced by Yolanda from the other side: 'if at some point something happens to me, a frost comes, and I have no more seed, some of the other guardians may have it and can provide it to me' [4]. Resilience through distributed seed ownership (a commons) and free circulation reinforces seed guardianship as a futural practice.

Guardians understand that no transition towards a regenerative or sustainable future is possible without seeds [3]. Guardianship of native and creole varieties thus becomes an important part of producing desirable futures. There is a widespread sentiment amongst guardians that AfroIndopeasant lifeworlds are under attack, and it is mainly operationalised through technologies and laws reducing possibilities for traditional seed practices.

Seed care and exchange are framed as a resistance to corporate dependence and a way to increase resilience to multidimensional Modern pressures. It is a pathway to food sovereignty beyond food security. Peasant guardian Yolanda ponders about this: 'if I have seeds and not only the money to go and buy the products, rather I have them, [I'll have] … healthy products that are clean' [4].

Recentring non-Modern institutions

As part of their work enacting alternatives to food production and exchange, seed guardians connect narratives and values associated with agroecology and ancestral institutions. Indeed, a common feature of seed guardians seems to be an interest in the recovery and revaluing of memory and traditional knowledge.

For Lina, there is a need to 're-educate ourselves again, claiming those traditional practices around custody of seeds, food, caring for water, caring for life, and caring for our health' [5]. As a neo-peasant who defected from urban life, Lina appreciates Indigenous and Afro-Colombian visions of respect towards nature and connection to the divine [5].

For Emiliano, the importance of ancestral agriculture is to advance transformations originating 'from ecosystems and ancestral production systems, … to recognise, differentiate, and define … agriculture [for particular regional conditions], and not everything being a single model' [30]. This 'rescue of memory' has allowed AfroIndopeasant seed guardians to move from having a vision of 'being stigmatised to … [visions of] being valued, being protagonists. [To overcome] … that underestimation, self-underestimation which they [experienced]' [2].

Indigenous agricultural institutions

Regretfully, going deep into each of the non-Modern institutions recentred by guardians is beyond the scope of this book. In this section, I will mention some of the institutions evoked by guardians and briefly define them as they were used in the context of the interviews.

Colombian neo-peasant Lina incorporates *pagamentos* into her daily activities as a guardian. In her words, *pagamentos* are

> offerings or payments … the fact of first asking for permission … When you enter a plot of land, when you are going to sow, when you arrive somewhere, you always ask for permission, and after … you also pay in some other way for what you're going to receive. It's … reciprocity [5].

Mexican guardian Rita mentions the *milpa*, an ancestral Meso-American cyclical cultivation method involving maize, beans, and squash, but more importantly, involving a sacred mode of relation between farmers, communities, crops, land, and the universe [31]. Ecuadorian guardian Pedro professes a 'deep love for the ancestral cultures of Ecuador' and mentions the Quichua terms *mallqui* and *sach'a* (respectively, agricultural fields cultivated by humans and 'untamed' nature) as interrelated poles of personal attraction [3]. Finally, Venezuelan guardian Alma frames the *conuco*, and a *conuquero* economy – that is, a system of shifting cultivation in smallholdings typical of peasant, family, and community agriculture – as a fundamental part of food system transformation [32].

Guardians like Ramiro and Carmen have worked to uncover the cosmovision of 'Andean Reciprocal Nurturing' (ARN).[10] For Ramiro, ARN distinguishes between 'old world' and 'new world' agricultural practices. It frames European plant domestication as a process of 'improvement towards greater productivity, selecting only the best plants [to reproduce] and discarding others' [1]. Conversely, ancestral Andean agriculture

> didn't discard the least productive plants because … living beings, water, air, mountains have a spirit. But also, those … other [plants] continue to be part of the family. Just as in a human family there are brothers [that] may be more active, more hard-working, … [but] everyone is loved equally … Those cousin species, grandmothers, or aunts are important to maintain biodiversity and are resources in case of a climate event [1].

ARN is seen not only as a contestation of productivist values but a recognition of interrelatedness and reciprocity, since 'just as we grow potatoes and animals, they also raise us' [1].

Alternatives vs. green revolution

Narratives of agroecology and other alternatives (e.g., regenerative agriculture) are also mobilised by guardians to challenge the green revolution and validate their practices. As explained by Emiliano, the green revolution model 'uses agrichemicals, improved seeds, [and an] expansion model [that was] designed to solve the problems of big industrialists [and] big companies, [without] really [helping or contributing] … to the problems of small farmers [and] families [living] in marginalised rural areas' [30].

By contrast, agroecology, organic farming, and other alternatives pursue 'healthy agriculture … without agrochemicals' [30]. Furthermore, they contemplate the 'entire production system … integrally. Do not focus on monocultures, [and] enable [farmers] to remain on their land, allowing them to live with dignity, making an appropriate use of … nature, [with] their own knowledge' [30].

For guardians, agroecological narratives are disadvantaged in relation to those of industrial food production, as the latter hold more power as they are mobilised by governments, universities, and private companies [33]. In fact, guardians go to great lengths to justify the validity of agroecology and ancestral agriculture. As Emiliano puts it,

> you have to demonstrate that there are communities that have been producing in a certain way for 10,000 years or 1,000 years, … eating something for so many … years and they haven't died, they're there, they're healthy, … the ecosystems are healthy [30].

Disadvantages also present themselves in unexpected ways, as consumers might prefer 'good looking' food over the diverse aesthetic of agroecological food products. Related to this matter, peasant guardian Yolanda reflects on her own experiences:

> [pesticide] residues remain in food [but] … the important thing [for the consumer] is that the lettuce is larger, the broccoli fatter. In my case … they're small, so there are people who say no, if the potatoes are too thin [they] don't like them. I have heard … that [some] weeks before taking out the product, [farmers] put an anabolic agent on it. I don't know exactly what it is … but it makes [food] look brighter, better looking, more attractive to people [4].

Despite these disadvantages and power asymmetries in food production narratives, Emiliano believes that alternatives 'must be strengthened to improve the living conditions of communities and small farmers' [30], especially as '[formal] education … prepares [people] … to adopt [the green revolution] model and expand it' [30].

Seed guardian futures

The big picture

Emiliano gives a succinct account that encompasses many of the shared values and future visions of seed guardians:

> We aspire to be able to develop diverse agricultural models that consider ancestral forms of production and allow life to adapt to changes taking place in ecosystems. That seeds circulate freely … considering that seeds are key to feeding people. That people continue to produce their food according to their conditions, not only environmental but also economic. That they have access to seeds regardless of their economic condition … That seeds are not monopolised by some companies to control territories and markets … That seed freedom is ensured [by] the system [30].

Other guardians use the broad vision of *Buen Vivir* as an organising principle for the future. Ramiro and his peers in the mountains of Colombia use the term *bien vivir* to guide their work, explaining that 'the goal is to nurture life, and [that] implies us flourishing, but that others and the territory also flourish … [In summary] it's the highest degree of happiness and fulfilment, … care, [and] defence of life systems' [1]. This sentiment is echoed by peasant guardian Yolanda, when she hopes that 'in the future everyone becomes aware and becomes more responsible with water, [and] … the forest, … a problem [that is not only local, but national, and global]' [4].

In Ecuador, visions of a regenerative future are prevalent and position seeds as essential [3]. This regenerative model refers to 'localised economies [and] localised production, not … moving food from here to there. [Food] with more diversity and higher nutritional quality, with more localised forms of governance' [3]. In this vision, guardians assume a Transmodern position when they exclaim that it

> doesn't mean that we return to an isolated primitivism, we are not primitivists! We must use the tools that modernity has already given us, the global village can continue to be connected, but be a global village, a set of autonomous communities that live in harmony with the environment [3].

Dignity, recognition, and networks

Guardians want a future where all farmers, and specially peasant farmers, can live a dignified life with the compensation they receive as food producers. For Rita, this might translate into new producer–consumer partnerships, where healthy food is prioritised, seeds are revalued, and both ends of the chain receive fair value. This is critical, because the peasantry is getting older and younger generations do not see this as an attractive life-path, as income from farming is increasingly precarious [31].

This dignified future is also a space where 'traditional knowledge, [non-hegemonic] organisations and seed custodians are recognised' [5]. As peasant guardian, Yolanda expresses it: 'I would hope that people become more aware that those of us who like to have a seed collection … it's not [because] we're weird' [4]. Beyond farmer-to-farmer relations, Lina hopes that the future brings more horizontal relations between technicians and farmers, dialogue, and knowledge exchanges that 'create spaces where those who lead are the communities, farmers, and seed guardians; people who have been working for generations on the issue of native and creole seeds' [5]. This is also expressed as a desire to strengthen seed guardian networks to 'be totally autonomous … groups of people working on issues of conservation, guardianship, recovery of seeds with super clear methodologies' [5].

Towards food sovereignty

Lina dreams of a future where 'people and organisations aiming at food sovereignty can be autonomous and self-sufficient' [5]. When speaking about the future, peasant guardian Yolanda refers to a recovery of peasant institutions that challenge food commodification. She actively complains about other farmers who

> have to go to town looking for a slice of onion at the supermarket because [they] don't grow them at home. [This is] unfortunate because one can have onions, coriander, parsley, everything one needs and have a clean and healthy self-nutrition [4].

This guardian would like a future where 'everyone dedicates a portion of [their] land, [their] orchard, to grow their own food and keep their own seeds' [4]. However, Yolanda is sceptical of this future, as people seem to think that supermarkets will always be fully stocked, and seedlings are now easily accessible commodities eroding even seedbed-making practices.

Scepticism is also present in the meaning of terms and who mobilises them. Concepts like sustainability, food security, food self-sufficiency, food sovereignty, or even agroecology are interpreted differently by incumbent actors [5], [31]. Lina is worried that '[policymakers] start using our terms, … misrepresenting concepts … created from peasant and social organisations for the benefit of large multinationals' [5]. On the other hand, Rita's concerns relate to governments and international cooperation agencies using these terms uncritically. Just as sustainability is used for greenwashing, phrases like 'national food security' or 'self-sufficiency' can advocate for food access without questioning who has the resources to produce it [31].

In this scenario, organisations working with these concepts may still exclude AfroIndopeasant practices or attempt to modernise them (i.e., industrialise, capitalise, make them more efficient, encourage use of agrichemicals, etc.). For Lina, the least eroded concept is 'food sovereignty', as it was created by peasant social movements, and she feels it is crucial that in the future this origin is fully recognised [5]. Food sovereignty includes '[producer] power over seeds, … biodiversity in peasant hands, the power of peasants to produce what they want, healthy food, [and] local markets' [31]. As such, it is 'totally against capital, and [transnational corporations] … can't just occupy it … lightly' [31].

Transition pathways

Although some guardians believe a deep crisis is needed to achieve the desired transformation, most guardians agree that their current practices aid in the construction of a more just and sustainable food system. In Mexico, guardians are developing conservation, characterisation, and participatory improvement projects, while implementing community seed funds[11] and seed fairs. They are also linking producers of native corn with buyers that might be able to offer fairer prices, with an ambition of ultimately differentiating native corn from hybrids, to achieve better pay for native varieties. Mexican networks also continue to advocate for government-enforced price guarantees and farmer subsidies [31].

In Venezuela, where a comprehensive seed law was passed in 2015 (see Chapter 5), guardians hope to see proposed institutions being established. These will be spaces for seed improvement, exchange, participatory certification, and mutual learning, based on a 'conjunction of peasant … and technical knowledge' [2]. Of these institutions, Carmen is most hopeful for the 'scientific-peasant alliance', which can aid in building community-managed germplasm banks [2].

For Alma, connecting diverse initiatives will be key for food system transformations. She expects to link old and new processes to strengthen the seed defence movement. For instance, connecting experiences of native seed guardianship with historical land struggles and consolidating government programmes like *Todas las Manos a la Siembra* (All Hands to Sowing) [32]. The Venezuelan process has been rocky since the passing of the law, due to the multidimensional crisis of the Bolivarian Revolution. However, 'things are still being done, advancing in ... small victories' [2].

Education and knowledge exchanges are a shared pathway towards food system transformations. For Alma, teachers and students should be involved with seeds from primary to university education, while Pedro proposes family-focused education using mass media, especially via the internet and social media [3], [32]. Strategies of inter-epistemic dialogue, or dialogue of knowledge(s) focused on AfroIndopeasant knowledge-holders, are also emphasised.

In Venezuela, community knowledge-holders have been officially named *maestres pueblo*[12] (people's teachers or competent knowers) who 'are now participating ... as structuring figures [at the micro-level]' [32]. Rita believes the focus should be on supporting and improving the dialogue process, since *maestres pueblo* 'are the ones who have the knowledge [and] experience' [31]. For Carmen, the ideal would be to institutionalise these dialogues so that 'the wealth of peasant knowledge [is] enhanced with techno-scientific interventions from a commitment and a vision of open transdisciplinary [mutual] aid' [2].

All these strategies surely advance food system transformations. However, as Pedro so aptly puts it, 'the most direct way to speed up this process would be for the state to listen, and ... apply these changes ... turning [them] into legislation' [3]. Processes of SGN–state interactions to transform system rules will be the subject of subsequent chapters. Three cases of national seed legislation change will be explored to understand the role of non-Modern practices and narratives in the Transmodernisation of food systems.

Guardians have dreamt of inclusive futures that value and enrich their practices (often framing them as utopian) but also recognise obstacles for their attainment. Rita and Carmen know the transition will be complicated, since 'the entire model needs to be rethought' [31] while we move towards scenarios of extreme climate change [2].

Potential blockages are often tied to industry and neoliberal policy pressures. For instance, Lina believes that government policy will always seek the benefit of large multinationals [5]. On the other hand, 'industry [has a] systemic incapacity to react to the problems the planet faces today, ... because for corporations, profit is the only objective' [3]. Pedro would rather not see industry as an antagonist in this transformation process but has not seen a 'single case where ... industry [has] limited itself [to] seek general wellbeing' [3]. Indeed, the plasticity and constant reconfiguration of agribusiness are seen as strong challenges to overcome [32].

Practices of Andean seed guardians

In the previous section, the narrative space of seed guardianship was explored to provide a map of meaning for their practices. In this section, the practices themselves will be described broadly, focusing on know-how, skills, and materials. This will provide an overview of seed guardianship moving from the individual and local community level to trans-local networks, and up to network–state interactions (See Figure 3.1). The focus of this section will be to provide a snapshot of current SGN practices. However, the evolution of these practices over time deserves more attention as their consolidation and diffusion aided the transformation of the whole system.

Although the focus of this book is to analyse the transformation space emerging from SGN–state interactions, the whole assemblage of practices described here should be seen as reshaping the system. As described in previous sections, seed guardianship has strengthened and reframed historical AfroIndopeasant practices over the last 25 years. This organisational process has created a novel innovative space where historically marginalised knowledge-holders are recognised and valued.

Starting at the individual and local community level, participants engage in seed guardianship, a practice described as the conservation of native, creole, and landrace seeds that are viable in a specific territory. Seeds are stored in a

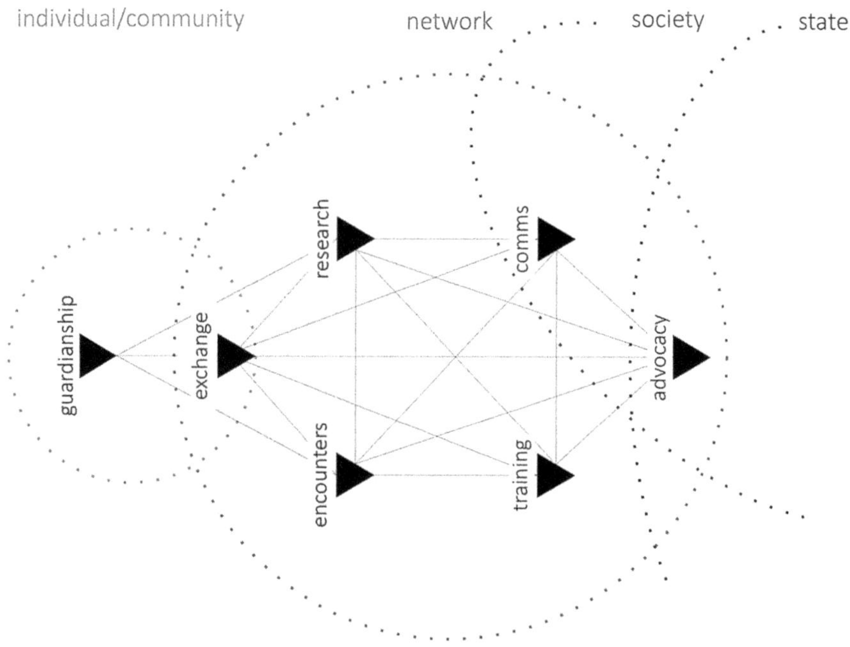

Figure 3.1 Practices of seed guardian networks across different scales

dedicated space, which can be part of a family home or a small community building. These seed houses or seed funds are made with locally available materials and provide shade and cool temperatures for seed preservation. Seeds may be stored in upcycled glass jars, small plastic bags, or clay pots vacuum-sealed with wax. All containers are labelled with seed names, and some guardians keep records of seed origin and other relevant data. However, seed houses do not store static collections; guardians must constantly (re)sow seeds to regenerate them and preserve their vitality.

At the local community level, but also expanding towards network interactions, seed exchange is done through gifting, barter, or commercialisation. Guardians may gift seeds to neighbours or other communities interested in growing different crops. They may also exchange seeds by bartering with other guardians or non-participant farmers who attend network events. Finally, guardians sell their seeds to other interested parties, sometimes with the aid of network databases. Contact also happens via social media or messaging apps, widening the geographical scope of trade, leading guardians to ship their seeds over long distances. Seed exchange is very active but is not without risks. Guardians mainly worry about contamination from industrialised or transgenic seeds, since there is no convenient way of determining the origin of seeds procured by strangers and crossbreeding is highly likely. Another major perceived risk is 'biopiracy', the appropriation of landrace, native, or creole seeds by corporate actors to use in private breeding programmes.

At the network level, seed guardians meet virtually and physically to share knowledge about seeds, discuss the current state of seed systems, organise, and engage in collective action. SGNs are fundamentally horizontal but are coordinated by small groups of committed participants aided by volunteers. These facilitators organise working groups or workstreams that attend to different dimensions of their mission, including:

- **Seed guardian encounters**: These are periodic events where participants can get together to exchange seeds and knowledge. These may be network-specific or attached to other agricultural-sector or community events. Guardians may sell, exchange, or barter seeds in these spaces.
- **Internal and external communication strategies**: Includes updating participant databases, organising virtual meetings, publishing virtually and physically on related topics (e.g., seed catalogues, books, reports, pamphlets, podcasts), developing consumer awareness campaigns, and managing internet presence with special emphasis on social media.
- **Participant training**: Includes participant empowerment, gender equality training, knowledge sharing on seed production and improvement, seed house conditioning, commercialisation, public policy, legislation, agroecology, risks of transgenics, amongst other things.
- **Research**: Includes the development of methodologies or programmes related to all other workstreams. Participatory seed improvement programmes and participatory guarantee systems for quality assurance are of special interest.

- **Advocacy**: Includes the development of strategies to resist legislation that undermines non-industrial practices, participating in lawsuits and other legal options to transform current laws, and pushing for recognition of practices by governments. This recognition can be exemplified by the establishment of National Corn Day (*Día nacional del maíz*) in Mexico or National Peasant Seed Day (*Día nacional de la semilla campesina*) in Venezuela.

Seed guardianship

Seed guardianship is a place-based practice in which one or several individuals from a community care for native, creole, or landrace seed varieties viable in the area. Arguably, the main skill of this practice is knowing how to maintain the vitality of seeds. As Yolanda puts it, 'rotating them above all, keeping them fresh, … the idea is that they go out to the field, go back and reproduce, germinate again' [4].

Maintaining vitality includes an embodied knowledge of place conditions that allow seeds to grow, such as soil, humidity, altitude, temperature, etc. For Lina, this is the first step of seed guardianship, to 'understand what seeds you can guard according to the conditions you have in the area' [5]. She elaborates, 'you begin to understand that it's not about guarding many seeds, but really the ones … that grow, the ones that you can keep in the area, keep them well' [5].

Maintaining seeds and 'keeping them fresh' also calls for knowledge on how to store them. Materially, guardianship requires a storage space and containers to hold different types of seeds. The space can be as simple as the corner of a room already in use at the guardian's home, but also as big as a dedicated small building serving a whole community. As an example, Yolanda's seed house is a small building without windows outside her house that served as her old kitchen. The walls are full of soot, and she calls a section of it 'the dark hole' since there is no lightbulb. Yolanda says that her more than 50 seed varieties are maintained by the climate of the building and justifies the absence of light since 'seeds like to be in the dark' [4].

These storage spaces have generally been dubbed by SGNs as seed houses, although in Mexico they are known as seed funds. As Rita admits, they are named as such to 'oppose [scientific seed] banks, [as] … in-situ seed funds [managed by] family or community … guarantee that there are seeds [in place]' [31].

Containers to hold seeds are widely varied. They can be upcycled glass jars with metal lids, small sealable plastic baggies, plastic produce containers, or clay jars vacuum-sealed with wax. There can also be no containers, as is the case of Yolanda's house, where maize is preserved according to Indigenous and peasant traditions, hanging from a horizontal rod near the ceiling. This is highly convenient for her, since 'birds help me a lot [spreading corn seeds to reproduce them], because the neighbours don't really plant that much corn'

[4]. This may be a signal that knowledge of preserving seed vitality is distributed between humans and non-humans, with material consequences, as the space enables animal participation in seed guardianship.

Containers appear also as part of family memories. As Lina recalls, 'I always saw jars of peach seeds in my grandparents' house.... My [peasant] grandmother, she means everything ... [she] was my teacher on ... recycling, storing, and saving seeds' [5]. Yolanda also uses upcycled glass jars. Her friend recently gave her a dozen, proof of community solidarity in guardianship. Yolanda always tells people 'that instead of throwing away [glass jars], they give them to [her]' [4]. She always disinfects jars with hot water before storing seeds and understands that 'if you leave [seeds] in a jar, they will lose their vitality' [4]. Rotating them to keep their freshness is important, 'so that they feed [the guardian] and along the way ... create new seeds to continue the process' [4]. Some guardians will also keep physical or virtual records of their seeds. This may be on a notebook, a personal spreadsheet, or sometimes using formats provided by SGNs. This effort is tied to seed exchange and research, as SGNs build databases of seed varieties, their characteristics, location, and who guards them. This allows for more fluid seed exchanges, knowledge-sharing, and opportunities for participatory seed improvement.

Seed exchange

Guardians exchange seeds with others guided by the belief that seeds should walk freely and as a strategy to increase resilience in the face of climate events. For Rita, these exchanges are acts of resistance against seed privatisation, transnational interests, and the imposition of rules [31]. Exchanges can occur directly between individuals or as part of organised events generally called seed fairs. Guardians often gift or barter seeds, exchanging them for other varieties of interest absent from their orchards.

As peasant guardian Yolanda puts it,

> when we meet in a barter, I show my seeds, so ... [they tell me] 'I don't have this one ... which one don't you have?' Then they begin to offer me, [and I tell them] 'I don't have that one, I'm interested' [4].

Barter plays a central role in seed exchange. However, there is an increasing interest in commercialising native seeds, as this could become a source of income to make the practice more financially sustainable.

Exchanges between individuals can occur at a local level between community members but are increasingly happening at a trans-local level due to the articulations of SGNs. Social media becomes the space for barter and commercialisation, as guardians make use of apps like WhatsApp, Telegram, and Facebook. Increase in use of these virtual tools is correlated with widespread adoption of smartphones around the region. Using these apps, guardians will either publish their seed stock or ask other network members for a specific variety.

Yolanda relays her experiences of digital exchange at the local level:

> Suddenly a neighbour tells me: 'I saw that in your [WhatsApp] status you published some corn, and I like it. I want some when we meet … and I'll bring you something else'. [So] I bring the [corn] and we'll meet in a part of town to [exchange it] [4].

At the network level, Yolanda also recalls one of her digital experiences:

> I … publish in the seed group, 'Who has X variety?' Then someone answers, 'I do, and I'm located [here]'. [They'll ask] how much quantity and the price [because there are people who sell them]. If I want them … I buy them, or exchange them if that person wants another seed instead [4].

In cases of individual seed exchange over long distances, guardians will send seeds by mail using courier companies. Seed exchange achieves national scale through social media use, phone calls, and mail trucks, as guardians send each other packages, paying for shipping at their end. Lina thinks these virtual exchanges exploded during the COVID-19 pandemic, as the social media pages of SGNs and participant organisations were bombarded with requests for native seeds [5].

This surge in demand provides new opportunities for commercialisation but also evidences gaps in knowledge about marketing [5]. As Lina describes, guardians 'prioritise … seeds for their own consumption, [then take] some for exchange, and lastly to market'. This is also one of the motivators for organising in networks that include isolated guardians, since marketing 'is fundamental [for] self-sustainability, and it's easier to do when you are associated or in a community than when you are a lone custodian … lost on a … mountain' [5].

Exchanges also happen as part of encounters organised by SGNs or other organisations working on food systems. These seed fairs may occur in public parks, community spaces, schools, or multisport courts of towns where prefabricated exhibition stands, tents, tarps, and desks are accommodated. Guardians may be accompanied by other food producers, sellers of cooked food and crafts, as well as local government authorities. For Lina, participating in these exchange meetings is an essential part of being a guardian, since 'you meet other people who are … working with your same seed varieties at other altitudes. You [encourage] an exchange of knowledge regarding … seed storage and conservation, and you … notice [how a] seed now has many varieties' [5].

Seed exchanges sometimes become a source for concern, mainly of contamination from pollination with hybrid or transgenic seeds, but also exhaustion of exchange networks. For Yolanda 'there are … many times where there is nothing more to exchange, [all guardians] already have the same seeds' [4]. Saturation will inevitably occur at some point, and guardians may feel uncertain about what happens when people cannot exchange varieties anymore. This is tied to the challenges of determining seed characteristics (i.e., seed quality) by people without access to laboratories, equipment, or scientific tests. At this point in time, there is no easy way for guardians to determine if two seeds of the same (or adjacent) variety are different from each other and how. Being able to test seeds and document this data (e.g., expected yields, water consumption, etc.) could better inform seed exchange in the future.

Related to this issue of unknown genetic composition, guardians are also concerned about farmers that come to exchange events with hybrid seeds to exchange for creole and native varieties. For Lina, generally peasants intuitively sow seeds they receive, putting native and creole varieties at risk. The solution would be to ask non-guardians

> to not bring [their own] seeds, if they want to obtain [native] seeds, [guardians] can provide them … because carrying seeds of unknown origin [is a risk]. That's an important pedagogical exercise that must be done about exchange to prevent this from happening [5].

Encounters

Seed exchanges and seed fairs are not the only encounters organised by SGNs. Most networks have institutionalised at least an annual meeting of seed guardians, there have been multiple international meetings, and there are many spaces for learning, research, and dialogue. Meetings were paused during the COVID-19 pandemic [31], with some being moved to virtual spaces. However, there has been a reactivation of meetings since the loosening of pandemic restrictions in the region.

National meetings are important for guardians as they gather a who's who of people operating in the same narrative and practice space. Pedro describes these annual meetings as 'an impressive thing … a time to meet the family, … people who think in a similar way [sharing] ideas' [3]. Some of these spaces have also been recognised by governments, as is the case of the annual National Peasant Seed Day in Venezuela [2], [32].

Research

Research led by SGNs might include seed characterisation, seed improvement, and reflections on their own practices. In Mexico, networks have published a catalogue with basic seed characterisation linking existing varieties to specific producers. The Ecuadorian seed guardian network also published a catalogue in 2018 showcasing nearly 3,000 preserved varieties. In Venezuela, networks developed an open seed information system hosted online[13] so that farmers could share information about their varieties, connect with others, and exchange knowledge.

These initiatives are seen as a way to protect native seeds and curve biopiracy, as there would be evidence of a history of seed ownership when a company wants to patent a seed type [31]. Deeper characterisation (i.e., genetic) is difficult, since SGNs do not have resources to fund them, and as some argue, 'every farmer who saves and selects his seeds generation after generation has his own variety … since characteristics [shift] depending on how they choose them' [31]. However, there are examples of genetic testing done in Colombia to identify transgenic material in Indigenous seeds using kits provided by the non-profit SWISSAID [5].

Participatory improvement programmes are also being developed by SGNs. These are spaces where farmers and scientists select seeds together according to the characteristics farmers want. As Rita evidences in the case of corn, peasant seed selection for improvement is faulty since

> many times they only choose the cob … that … looks very nice, big, … how they want it. But if [the plant] was on the edge of the plot, then it doesn't have as much competition with other plants, so we don't know if it's genetically a good plant [31].

Network participants may also make use of government programmes to improve their seeds, though not all guardians approve of this approach [30]. In Colombia, the seed authority has been implementing a national seed plan for this purpose, and in Venezuela, the 'All Hands to Sowing' programme seeks to involve citizens at all levels in food production.

The networks also publish constantly about their positions, work, and progress in their agendas. There are countless blogs and websites where SGNs share long-format articles. Colombian SGNs have a virtual magazine where participants, activists, and academics share their views on native seed conservation. In Venezuela, research has also been done from within the movements reflecting on the legislative transformation process. This was published as a virtual open-access book titled *Semillas del pueblo* (*Seeds of the people*) [34] in collaboration with government, NGOs, and the social movements involved [32].

Communications

Communication strategies are important for SGNs since they interact with multiple actors in food systems and beyond. Communication occurs at multiple levels: guardian to guardian, within the network (e.g., between network coordinators and allied organisations), between the network and government actors (with both allies and antagonists), and with broader society.

Some of the networks even started as communication strategies, as it is the case of Ecuador, where founders went to communities and

> talked about how seeds were being lost, … the dangers of contemporary seeds, and then asked people about seeds they had lost. [They] made lists of lost seeds and of people … who may have them in the area, and people got very excited [3].

Communication between guardians may happen over the phone, chats, or social media apps. Meetings are also generally virtual, using platforms like Zoom, Teams, Facebook, or Skype. Guardians admit that virtual meetings were somewhat exclusionary in their first years of operation, since not all members had access to technology. However, this has improved as mobile phone adoption increased, and more tools are available [5].

Due to the diversity of people and organisations participating in SGNs, their interactions are inevitably intercultural. However, this is not immediately evident for guardians when prompted in interviews. As Pedro puts it, 'there was no pretence of interculturality at the beginning' [3]. For him, guardian-to-guardian dialogues in the social process of SGNs occur under conditions of equality. Furthermore, interculturality is not an empty category as used by other organisations pretending to embellish their processes. Pedro argues that 'in the network we do feel this intercultural dialogue causes real changes in people's lives … [from] the wealth of the other' [3].

Regarding inter-organisational dialogues, spaces for discussion were promoted by several SGN members over the years, resulting in more flexible and horizontal governance structures. One such example is the 'Alliance for Biodiversity' in Colombia, which was essential in establishing dialogues between organisations in defence of native seeds, including government, NGOs, academia, social organisations, and seed guardians [30]. These spaces for dialogue also include national and regional seed guardian encounters, academic forums, and participation in food sector events.

Analysing the wider process of articulation of SGNs and seed sovereignty social movements in the region is beyond the scope of this book. However, it is clear that international connections 'strengthen the struggles and agendas of the social organisation' [32]. As mentioned in a previous section, there are many actors around the world working on regenerative agriculture, food and seed sovereignty, and other alternatives to industrial food production. These actors engage in constant feedback loops. This international cross-pollination is seen as an important but generally unexplored aspect of seed guardianship consolidation.

Transnational information flows manifest as success stories that inform transformation, failures that evidence risks of inaction, or support for the agendas of national seed movements. Alma provides examples of all three regarding the Venezuelan process. First, at a local level, bringing information about packaged Peruvian native potato chips being sold as snacks at the airport allowed Venezuelan native potato farmers to see 'a diverse world of possibilities, ... contrary to what [incumbents] led them to believe, [a bag of chips becomes] a wonderful thing that uncovers magic' [32]. Second, the Colombian documentary *9.70* [35] dramatised the risk of inaction in the face of 'anti-peasant legislation' that was being enforced throughout the region; likewise, the experience of Paraguay with transgenics was used as a scenario to avoid.

Regarding SGN interactions with the broader public, guardians see information flows from rural areas into cities as fundamental. These can take the form of consumer awareness campaigns on the value of agrobiodiversity and cultural biodiversity [31], the health risks of industrial agriculture and benefits of agroecological and organic farming, or alliances with chefs to promote the flavour, colour, and nutritional quality of native food products [30]. Some of these campaigns have been very successful, being eventually institutionalised by national governments. Such is the case of Mexico, where the *Sin maíz no hay país* campaign turned National Corn Day into an official celebration in 2019 [31]; or Venezuela, where National Peasant Seed Day is celebrated yearly with state support [2], [32].

SGNs are actively publishing their experiences to reach increasingly broader audiences. Seed catalogues, books, academic publications, websites, and virtual magazines are just some of the products advancing their cause. SGN coordinators are always on the lookout for ways to communicate with citizens and sway public opinion in their favour. A particularly successful case of audience outreach occurs in Ecuador, where the network developed the web-based educational platform *Madresemilla* (Mother-seed) with access to low-cost massive open online courses, and the podcast *Radio Semilla*, where native seeds and other topics around transformation to sustainable futures are discussed [3].

Since 2018, the seed guardian network of Ecuador has developed a process of information dissemination to the public via the internet. After employing a community manager, their Facebook posts have reached more than 1.4 million people per year, their podcast has more than 8,000 listeners, and a virtual festival organised during the pandemic lockdown attracted 18,000 people from all over Latin America [3].

As has been mentioned previously, the production of information is also asymmetrical in food systems. SGNs are actively generating 'positive and favourable information so that public policies go towards these new alternative models, ... [and towards] native and creole seeds' [30]. However, building up information showing the successes and benefits of agroecology, AfroIndopeasant agriculture, and native seeds is not easy. Neither SGNs nor participating NGOs and grassroots organisations have the resources to fund appropriate research, while private companies spend millions developing their technologies and marketing strategies [30]. Producing successful examples and making them visible to broader audiences is a central ambition of SGNs and it is crucial in positive network–state interactions as described in subsequent chapters.

Information technologies

Many guardians emphasise the relevance of information technologies in their practices, particularly since the COVID-19 pandemic [2], [3], [4], [32]. Of these digital spaces, the main point of encounter is the WhatsApp chat. This app allows participants to be 'permanently connected, [and] … always in communication' [32], especially in Venezuela, where travel is difficult due to fuel scarcity.

The chat allows for broad participation across regions, biological phenotypes, ethnicities, and age groups, even leading to surprising connections, as was the case of Venezuela, where the Minister of Urban Agriculture is connected via WhatsApp. The chat is a space for sharing information, consulting, training, organising, exchanging seeds, connecting to international initiatives, and even holding events like the 2020 Venezuelan National Peasant Seed Day. As Pedro describes it:

> we have a chat on one of these services and … for three years it has been the most intense, most active form of bonding. Everything circulates there, always focused on topics of interest to the network. It's not a space for proselytising, or jokes, or talking about football, or anything like that. But hey! You're always going to see photos of mushrooms there … that's typical. [People will ask] 'Which mushroom is this?' We have like 4 or 5 mycologists … every time, [they'll ask] 'is it edible?' Photos of seeds obviously, plants.… 'What plant is this? Do you eat this plant or not?' Photos of pests and diseases. 'How do I control it?' Photos of wonderful insects and all these things continually flow. 'How do I sow this or that? How do I cook it?' It's a lot of information being exchanged. Sometimes it makes me sad that we don't copy everything because these discussions … are wonderful [3].

Advocacy

Seed guardianship, exchange, and networking practices were ultimately exhausted as transformative forces, leading SGNs to engage in politics to pursue the world they want. As neo-peasant guardian Lina puts it: 'I began to understand that the issue of seeds was not only limited to being in one's territory guarding them, but also that … political advocacy was very important to transform … [the] reality of our surroundings' [5]. This is not to say that network practices did not play a role in system transformation; in fact, the political mobilisation of guardians is deeply related to internal empowerment connected to their daily practices.

Decades of work by the networks led to the 'formation of a … new self-esteem, gained by [their] seeds and [their] knowledge, defending and [ultimately] asking for [legislative change]' [2]. Revaluing AfroIndopeasant practices, strengthening them, and connecting outliers that enact them led to more assertive participation. As Carmen explains of the Venezuelan legislative process, '[guardians] participated without fear … saying, "yes, yes, yes. The peasantry *is* the guardian of seeds! *yes!* we produce seeds, and we do it *this* way" [emphasis added]' [2].

As mentioned in previous sections, SGNs are deeply distrustful of the state, as they see governments colluding with corporations to create a hostile environment for native seed circulation. For some guardians, corruption at the state level is rife and civil society generally cannot participate in decision-making. Related to this position, there are network participants who are actively anti-state, closed to dialogue with government or state-led interventions that may favour SGN agendas [31].

This produces a conflicting attitude in SGNs; despite state agro-industrial biases, neoliberal leanings, and a historical invisibilisation of their practices, networks know that they must interact with national governments to preserve their way of life. These network–state interactions are always fragile, as SGNs have been generally defensive and believe they 'should never lower [their] guard' [5]. However, in recent years the networks have shifted from a defensive position to the offensive, as they moved to building concrete proposals to transform national seed systems.

Ultimately, SGNs advocate for recognition of their practices and narratives, and the creation of a set of rules that levels the playing field for all seed system actors. To achieve this, political advocacy activities mobilised by the networks include dissenting, campaigning, advising, monitoring and pressuring governments, advancing legal challenges to existing and emerging legislation, opening spaces for dialogue, and proving validity of their practices.

In most cases, network advocacy is not unidirectional, as the state has also opened spaces for dialogue. Despite the generally pessimistic views SGNs hold of government, the state is not a monolithic enemy. Network participants are still aware that many government actors have been previously employed by companies like Monsanto or Syngenta, so caution persists. However, SGNs always find government allies that sympathise with their narratives and facilitate SGN manoeuvring. State allies will leak legislative drafts to alert SGNs of upcoming changes [2], [3], [32], and even in Colombia (arguably the most recalcitrant right-wing country of those included in this research) state actors have started including SGNs in their discussions over the years.

It is important to emphasise that left-leaning governments have provided more spaces for social movement participation in decision-making. The impact of the Latin American pink tide[14] cannot be overstated. In Venezuela, framing the state as eco-socialist was crucial to the development of the 2015 seed law, which recognises AfroIndopeasant practices and native seeds. In Ecuador, a *Buen Vivir*-centric state opened up spaces like the Plurinational Intercultural Conference on Food Sovereignty (COPISA), where SGNs and other social organisations drafted a seed law (see subsequent chapters).

As Rita reflects on the experience in Mexico,

> [a leftist president] … created a different panorama for us, suddenly many who were in the *Sin maíz no hay país* campaign were in government [positions] … In 2018–19 we found ourselves in a situation where [government] began to ask us: 'What do you want? How do we protect [native] corn?' … suddenly that's super difficult, right? [To] come up with grounded proposals [31].

Advocacy is deeply tied to communications, so SGN and SGN-adjacent campaigns are constantly mobilised to influence legislative outcomes. These campaigns might be for or against existing or upcoming legislation. Anti-transgenic campaigns have been generally influential in all countries, with the GMO-free Venezuela being arguably the most influential at the national level [32]. Campaigns designed to protect specific food staples have also garnered support for seed guardianship, such as the case of Mexico, where *Sin maíz no hay país* and *Alianza por nuestra tortilla* (Alliance for our tortilla) have put native seeds at the centre of public discussion [31].

Seed-specific campaigns like the Venezuelan National Peasant Seed Day or the Colombian Seeds of Identity (*Semillas de Identidad*) were the starting points for the creation of SGNs in those countries. SGN-adjacent campaigns against free trade agreements, and for land reform, farmers' rights, water protection, among other issues, also aid the advocacy work of the networks. An important example of a communications piece having transnational impact is the Colombian documentary *9.70*. This film spread the message of alleged anti-peasant legislation enforcement throughout the region, informing multiple transformation processes.

Lawsuits are one of the most widespread tools in the advocacy arsenal of SGNs. In Mexico, the networks have filed lawsuits against transgenic corn, leading to a moratorium. In Ecuador, SGNs have campaigned against multiple laws that affect their work, most notably to rewrite 15 articles of the national seed law. Interestingly, the Ecuadorian seed guardian network sued the Ministry of Agriculture to release information about the regulations they were drafting. As Pedro frames it, 'I had to sue them to be able to speak to … [engage in] dialogue with the Ministry of Agriculture, while big capitalist companies call the minister directly every week' [3]. Dissidence and resistance are constants in SGN advocacy, whether against neoliberalism, the green revolution, capitalism, concentration of land ownership, glyphosate, or over-exploitation of resources.

SGNs not only resist legislation but also contribute by developing proposals and supporting initiatives that advance their agendas. In Mexico, the networks have actively supported policies of price guarantee for food products overseen by SEGALMEX, the body responsible for food security [31]. The Ecuadorian seed guardian network participated as an advisor to the National Assembly in drafting the 2008 Constitution [3]. In Venezuela, SGNs were highly active in the development process for the 2015 seed law: they gathered signatures in support of their interests, monitored what was happening in the National Assembly, organised a process of consultation (popular debate workshops), and drafted a proposal for the law [2], [32]. In Mexico, SGNs have developed a draft of the Law for the Promotion and Protection of Native Corn, which includes the creation of a Corn Council that envisages mandatory participation from civil society, Indigenous peoples, rural communities, and peasants [31]. And in Colombia, there have been joint projects and an ongoing co-visioning process for the definition of a national seed system [30], as well as participation in policy guidelines for agroecological agriculture.

Knowledge-wise, SGN advocacy practices remain a mystery; there are no mentions of lawyers, politicians, legislators, or political scientists in their ranks. Advocacy appears to also respond to cycles of action–reflection–action, as they are citizens participating in decision-making from an outside position. As Pedro puts it,

> we are always out, we appear suddenly, we do a lot of things and again we disappear … I feel that we are seen by other actors as excessively informal, radical. And I do feel that some allies … are often annoyed seeing the network's management and outreach capacity, because we don't move within the normal schemes in which all these movements move [3].

Despite their unorthodox approach, SGNs have become a powerful force in seed systems. Industry and transnational corporations almost have guaranteed seats at the table, while civil society must open spaces to have a say in seed legislation. But as Rita explains, 'there are little entrances, like small fractures, where you can enter and make them bigger, and create the world that is in your [interest]' [31].

It is necessary to mention again the burden of proof as part of knowledge associated with advocacy. As Emiliano puts it, 'the issue of advocacy in public policy forces us to demonstrate, … to influence policy … we … must show those who have power … that [our practices are] good and that [they] work' [30]. This requires skills, both to develop successful experiments with demonstrable results, and communicate them to an agnostic audience. Part of the advocacy work of SGNs is to demonstrate the benefits of agroecology, organic farming, AfroIndopeasant agriculture, and native seeds.

Emiliano is deeply troubled by this. As he expresses it:

> this seems absurd to me. Demonstrate that organic farming is good for the environment? When have those who are producing damage with agrochemicals, with that model, had to prove anything? They are the ones who must show what's good, they're the ones who have to show that transgenic seeds do no harm. And they put the burden of proof on us. [It's] the world upside down! The one who goes with the gun must prove

that the gun doesn't kill. So, [we] have to [make] things … right. Those who must show that they do no harm are the ones who come up with new things. This is a very unequal dilemma in … public policy [30].

Ultimately, the objective of their advocacy is to develop public policy that allows for informal(ised) and formal seed systems to coexist. This is first and foremost an exercise in visibilisation. As Emiliano puts it, 'the [informal] system works, [but] it isn't visible, characterised, it isn't written, there's no [official or institutional] document … So what you have to do is make it visible, show it' [30]. This is a titanic endeavour, since there are pervasive agro-industrial biases at the state level, and the neoliberal state's strength is constantly deployed to defend corporate interests. Furthermore, SGNs are going against 500-plus years of Coloniality, which has pushed their narratives and practices to the periphery.

As Carmen explains, hegemonic actors speak of 'Afro-descendant seeds [as] … linked to blackness, to slavery, to backwardness' [2]. Such is the challenge of pursuing a Transmodern decolonial future while dismantling colonial structures, worldviews, values, etc. Even if the process of creating an egalitarian set of rules is successful, a law that enables the coexistence of industrial and AfroIndopeasant practices may be full of contradictions. How can you draw clear physical boundaries between transgenic and native seeds? How would a law stop both from sharing genetic material? How can regulation ensure that airborne agrichemicals don't spread to communities and crops where they are not wanted? How can the state support AfroIndopeasant ways of being without aligning them with Modern values or practices (i.e., without handouts of hybrid seeds and fertilisers)? And ultimately, who has control of national agri-food patrimony moving forward?

Conclusion

Latin American seed guardian networks have consolidated over the last 25 years, ultimately shaping seed legislation at a national level. This organisational process has been first and foremost an exercise in revaluing and reframing existing seed production and exchange practices typical of AfroIndopeasant regimes. However, not all Indigenous, Afro-Latino, or peasant farmers are seed guardians; this category refers to a special type of knowledge-holder with a particular skillset.

Guardians are deeply curious and experimental; they are prefiguring seed traits in their fields but also futures of food sovereignty and autonomy. Seed guardianship is a place-based practice that quickly transcended the local level to become trans-local with presence at national, regional, and global levels. This organisational process of seed guardian networking has been a simultaneous creation of a shared narrative field and the consolidation/diffusion of practices.

SGNs have created a flexible and constantly shifting discourse coalition, a narrative field that both coalesces the network through shared purpose and pulls external allies into its orbit. It differentiates native, creole, and landrace seeds in AfroIndopeasant hands from those of industry. It paints a picture of high control over food systems leading to increasing dependence, and networked efforts of seed custody as a pathway to resilience. It reframes and revalues Indigenous agriculture and other alternatives against the failure and persistent damages of the green revolution. It also dares to dream of transition pathways to alternative futures where plural agricultural self-sufficiency models are locally developed to preserve the dignity, autonomy, and freedom of peoples and seeds. This narrative field has provided meaning to seed guardianship practices, strengthened practitioners by reframing traditional seed care as futural, and configured guardians and their allies as a coalition that shares narratives.

Through their interactions, seed guardians have experimented with introducing and interchanging new meanings, materials, and competences. This has simultaneously Transmodernised traditional seed care practices and the rules that govern national seed systems. Recognising and renaming the practice of seed guardianship, connecting practicing outliers, creating seed houses, experimenting with education and training providing new skills, or with materials such as genetic-testing kits or virtual databases, have all contributed to the emergence of an influential new sub-system. The consolidation of these practices has effectively created a niche space that innovates according to Transmodern values and visions. The consolidation of SGNs has made the previously invisible visible; it has broken the abyssal line, showing hegemonic actors that which they ignore and informalise.

Narratives and practices of SGNs position them as a sustainable materialist social movement, a political organisation based in sustainable material practices, collective action, and prefigurative politics [36]. SGNs emerged as a response to the unsustainability and injustices of the dominant global food system. They are a Transmodern reaction to the exhaustion of Modernity/Coloniality, embodying non-Modern narratives and practices that unsettle both everyday material flows and high-level structures.

Ultimately, guardians realised that system rules had to be changed in order to guarantee their permanence and build a transition to their desired futures. Through advocacy, network–state interactions have opened a translational contact zone where science and law are unsettled, allowing Transmodernity to emerge. Chapters 4–6 explore cases of translational contact zones opening in Colombia, Ecuador, and Venezuela. They focus on network–state interactions that attempted (with varied degrees of success) to Transmodernise system rules with SGNs as a protagonist transformative force.

Notes

1 The term *gringo* is a colloquial way of referring to foreigners, especially people from the United States of America.

2 Also known as seed custodians.
3 Referring to agro-industrial 'terminator' seeds, which inhibit reproduction after one cycle.
4 The term neo-peasant refers to urban dwellers who decide to move to rural areas to pursue a life dedicated to food production, small community life, and subsistence.
5 The Spanish term *compadre* is generally used to refer to the godfather of a child, but it is also used as a term of endearment for very close, almost familial, friendship relations.
6 For more details on each of these concepts see [8], [12]. For an extensive map of international agroecological initiatives see [17]. For a different taxonomy of social movements responding to food inequalities see [13].
7 For more visit https://viacampesina.org/en/
8 For a more nuanced history of seed activism see [22].
9 https://navdanyainternational.org.
10 This term is the author's translation of the Spanish phrase *crianza mutua Andina.*
11 The Mexican term 'seed fund' refers to local and community-run germplasm banks. Networks in other Latin American countries refer to them as 'seed houses'.
12 The Spanish phrase *maestres pueblo* (popular teachers or competent knowers) refers to local knowledge-holders, particularly those community experts with experience of local seed varieties. In the legislation, they are referred to in the masculine as *maestros pueblo*, but here the author chooses to use the genderless *maestres.*
13 The Venezuelan information system was initially hosted in the domain www.elconuco.com.ve, now offline.
14 The pink tide refers to a wave of left-leaning political parties winning presidential elections in South and Central America at the beginning of the millennium.

References

[1] NET001, 'private communication', Feb.2021.
[2] NET008, 'private communication', Oct.2021.
[3] NET006, 'private communication', Nov.2021.
[4] NET002, 'private communication', Oct.2022.
[5] NET003, 'private communication', Mar.2021.
[6] ETC Group, 'Who will feed us?' 2017. Accessed: 23 Nov. 2025. [Online]. Available: https://www.etcgroup.org/whowillfeedus.
[7] P. A. Loring, 'Regenerative food systems and the conservation of change', *Agric Human Values*, vol. 39, no. 2, pp. 701–713, Jun.2022, doi:10.1007/s10460-021-10282-2.
[8] A. Bless, F. Davila, and R. Plant, 'A genealogy of sustainable agriculture narratives: Implications for the transformative potential of regenerative agriculture', *Agric Human Values*, vol. 40, no. 4, pp. 1379–1397, Dec.2023, doi:10.1007/s10460-023-10444-4.
[9] M. D. Anderson and M. Rivera-Ferre, 'Food system narratives to end hunger: Extractive versus regenerative', *Current Opinion in Environmental Sustainability*, vol. *49*, pp. 18–25, 2021, doi:10.1016/j.cosust.2020.12.002.
[10] P. Webb, T. G. Benton, J. Beddington, D. Flynn, N. M. Kelly, and S. M.Thomas, 'The urgency of food system transformation is now irrefutable', *Nature Food*, vol. 1, no. 10, pp. 584–585. doi:10.1038/s43016-020-00161-0.
[11] S. J. Buckton *et al.*, 'Transformative action towards regenerative food systems: A large-scale case study', *PLOS Sustainability and Transformation*, vol. 3, 11 November2024, doi:10.1371/journal.pstr.0000134.

[12] B. P. Oberč and A. A. Schnell, 'Exploring the pathways towards the future of farming Approaches to sustainable agriculture', 2020. [Online]. Available: https://twitter.com/IUCN/.
[13] R. Motta, 'Social movements as agents of change: Fighting intersectional food inequalities, building food as webs of life', *Sociological Review*, vol. 69, no. 3, pp. 603–625, May2021, doi:10.1177/00380261211009061.
[14] K. A. Congreves, 'Regenerative agriculture – a definition and philosophy', *npj Sustainable Agriculture*, vol. 3, no. 1, p. 60, Nov.2025, doi:10.1038/s44264-025-00097-7.
[15] S. Narula, 'Beyond reform: Food sovereignty and the future of global food systems', *Indiana Journal of Global Legal Studies*, pp. 141–203, 2024. [Online]. Available: https://muse.jhu.edu/article/949739.
[16] L. Ashley, D. M. Rodríguez Herrera, and A. Hummel, 'Seed saving and seed sharing: Occupying a liminal legal space', *Studia Socjologiczne*, vol. 2025, no. 2, pp. 191–216, 2025, doi:10.24425/sts.2025.154902.
[17] O. F. Giraldo, 'The agroecological movement. A panoramic view', *Journal of Peasant Studies*, vol. 52, no. 1, pp. 47–73, 2025, doi:10.1080/03066150.2024.2318463.
[18] D. Cruz and E. Van De Fliert, 'The ethics of food sovereignty: Discourses for transformative social change and community development practices by peasant movements', *Community Dev J*, vol. 58, no. 1, pp. 64–78, Jan.2023, doi:10.1093/cdj/bsac034.
[19] R. M. Hansell, 'Indigenous food sovereignty: Literature review', *Fourth World Journal*, vol. 24, no. 2, pp. 147–184, Aug.2025, doi:10.63428/7zjhvq86.
[20] J. Kloppenburg, 'Seeds of struggle: Plant genetic resources in the world system', in *First the Seed: The Political Economy of Plant Biotechnology*, 2nd ed., 2004. [Online]. Available: https://www.ebsco.com/terms-of-use.
[21] O. T. Coomes *et al.*, 'Farmer seed networks make a limited contribution to agriculture? Four common misconceptions', *Food Policy*, vol. 56, pp. 41–50, Oct.2015, doi:10.1016/j.foodpol.2015.07.008.
[22] K. Peschard and S. Randeria, '"Keeping seeds in our hands": The rise of seed activism', *Journal of Peasant Studies*, vol. 47, no. 4, pp. 613–647, 2020, doi:10.1080/03066150.2020.1753705.
[23] J. Kloppenburg, 'Re-purposing the master's tools: The Open Source Seed Initiative and the struggle for seed sovereignty', *Journal of Peasant Studies*, vol. 41, no. 6, pp. 1225–1246, 2014, doi:10.1080/03066150.2013.875897.
[24] C. I. Sánchez Gómez *et al.*, *Milpa corazón: las milpas de los guardianes.* Mexico City: Comisión Nacional para el Conocimiento y Uso de la Biodiversidad (CONABIO), 2022. Accessed: 24 Nov. 2025. [Online]. Available: https://bioteca.biodiversidad.gob.mx/janium-bin/detalle.pl?Id=20251125140515.
[25] E. Herrighty and C. G. Hill, 'The seeds are coming home: A rising movement for Indigenous seed rematriation in the United States', *Agric Human Values*, vol. 41, no. 3, pp. 1007–1018, Sep.2024, doi:10.1007/s10460-023-10532-5.
[26] L. J. Nikol, C. Almekinders, and K. Jansen, 'Seed activism on four fronts: MASIPAG's rice seed struggles in the Philippines', *Agric Human Values*, vol. 42, no. 3, pp. 1977–1995, Sep.2025, doi:10.1007/s10460-025-10747-8.
[27] V. García López, O. F. Giraldo, H. Morales, P. M. Rosset, and J. M. Duarte, 'Seed sovereignty and agroecological scaling: Two cases of seed recovery, conservation, and defense in Colombia', *Agroecology and Sustainable Food Systems*, vol. 43, no. 7–8, pp. 827–847, 2019, doi:10.1080/21683565.2019.1578720.
[28] S. B. Cohn, 'Lending seeds, growing justice: Seed lending in public and academic libraries', *Library Quarterly*, vol. 94, no. 2, pp. 117–133, Apr.2024, doi:10.1086/729231.

[29] P. Singh, 'Politics of knowledge in development: Explorations in seed sovereignty', *Studies in Indian Politics*, vol. 9, no. 1, pp. 105–117, Jun.2021, doi:10.1177/2321023021999179.
[30] NET009, 'private communication', Jan.2022.
[31] NET005, 'private communication', Apr.2021.
[32] NET007, 'private communication', Oct.2021.
[33] NET001, 'private communication', Oct.2022.
[34] M. Pérez *et al.*, Eds., *Semillas del pueblo.* Caracas: Editorial La Estrella Roja, 2016.
[35] V. Solano, Dir. *9.70*, 2013. [Film].
[36] D. Schlosberg, 'From postmaterialism to sustainable materialism: The environmental politics of practice-based movements', *Env Polit*, vol. 34, no. 1, pp. 92–112, Jan.2025, doi:10.1080/09644016.2019.1587215.

4 Colombian seed system transformations

In Colombia, AfroIndopeasant actors have been historically excluded from decision-making as their practices are routinely ignored or even suppressed by incumbents. In recent years, however, a combination of bottom-up actions and national-level inclusion efforts have resulted in a participatory approach to transform the Colombian seed system.

This exploratory process aims to conceptualise a national seed system for Colombia and was officially launched in 2019 by AGROSAVIA (Colombian Corporation for Agricultural Research, previously known as CORPOICA). Crucial to this initiative, and radically different from previous government-led efforts to transform the national seed sector, is the inclusion of the Organización Nacional Indígena de Colombia (ONIC, National Indigenous Organization of Colombia) and seed guardian networks (SGNs). Parallel to this process, a series of policies and policy guidelines emerge recognising the importance of native seeds and small-scale family farming.

This chapter explores the events that led to changes in the Colombian seed system during the last 30 years, presenting them chronologically as precedents of current efforts to conceptualise a novel seed system. It presents the recent interactions between AGROSAVIA and seed guardian networks to create seed system visions after more than a decade of tensions between two opposing but overlapping ways of producing food (industrial and AfroIndopeasant), and their corresponding worldviews (see Figure 4.1).

As presented in Chapter 1, these tensions have their origin in colonial processes and are maintained by the interrelated systems of power of Coloniality. Although actors do not reference their actions as decolonial, the process described in this chapter touches upon issues related to Coloniality and evidences the emergence of pathways to Transmodern futures.

In the first decade of this century, Colombia experienced an intensification of neoliberalism in the form of international trade partnership agreements. Liberalisation of the economy expanded the power of multinational seed corporations, aligned the country with international seed standards and regulations, and allowed the introduction of transgenic seeds. In parallel, Indigenous communities, in cooperation with international non-profits like SWISSAID, declared their territories transgenic-free. This created an opening

DOI: 10.4324/9781003715092-7

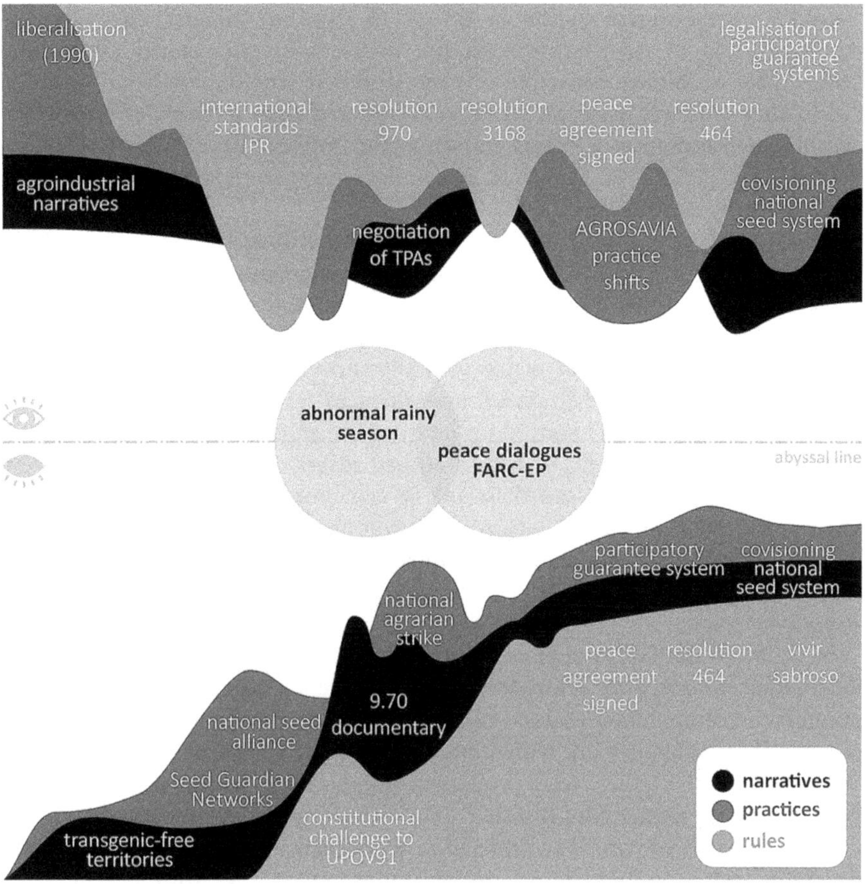

Figure 4.1 A visualisation of Colombian seed system transformations as a result of bottom-up and top-down pressures

for the creation of seed guardianship networks aligned with the international defence of native seeds.

For some authors, structural obstacles for participation in the legal regime led farmers and seed movements to focus on seed saving as a strategy for transformation [1]. And although these social movements initially relied on international NGO discourses to develop their tactics, they matured to put more weight on contextualised community needs [2].

In 2010, an abnormal rainy season wrought havoc in rural Colombia, leading to a multidimensional crisis for which the government offered an insufficient response. That same year, the controversial Resolution 970 was published, a confusing piece of legislation that planted the idea of state criminalisation of AfroIndopeasant seeds and related practices. This was exacerbated by a narrative shock to the system in the form of an independent documentary published on YouTube, titled *9.70* [3].

Lack of government action in favour of rural actors and narratives of criminalisation of AfroIndopeasant practices led to the consolidation of a novel social movement grouping diverse rural concerns that went on strike in 2013. That national agrarian strike achieved a change in seed legislation by eliminating Resolution 970 and clarifying the applicability of new legislation for industrial seeds only. However, AfroIndopeasant seeds were kept in a non-legal space by state actors.

These changes occurred as the negotiations between the Colombian government and the FARC-EP guerrillas were progressing. In 2016, a peace accord was signed that prioritised structural rural reform in the country while timidly pointing towards *Buen Vivir* transdevelopment framings. The peace accord between the longest active guerrilla movement in the world and the Colombian government had direct consequences on agricultural policy.

In 2017, CORPOICA[1] started an internal transformation that would allow it to handle the challenges of a post-accord rurality. A culture of collaboration was emphasised, a seed department was created, and the organisation appeared to be experiencing a turn towards more involvement from the social sciences. Parallel to this shift, the Ministry of Agriculture developed novel legal frameworks for non-industrial agriculture that appeared to promote *Buen Vivir* values.

In 2018, CORPOICA rebranded itself as AGROSAVIA in an attempt to rehabilitate its image as a top-down research agency disconnected from rural populations, a reputation that led to deep mistrust from farmers throughout the country. Regarding its work with seeds, AGROSAVIA implemented a National Seed Plan to strengthen seed producing organisations and, in 2019, started bringing seed system actors together to co-create a national seed system.

This visioning process was done from a complex systems perspective, purposefully including Indigenous organisations and SGNs with openness to their worldviews and practices. Two crucial aspects differentiate this process from previous attempts at conceptualising a national seed system: (i) recognition of peasant system potential for quality seed provisioning that responds to local environments and is fairly priced; and (ii) valuing all knowledge(s) involved in production, collection, characterisation, conservation, and use of agrobiodiversity [4].

2000–2010: Seed guardian networks and neoliberalism

The emergence of SGNs can be traced to the Campaña Semillas de Identidad (Seeds of Identity Campaign), a programme created to recover traditional seeds with Indigenous and peasant producer organisations. It was established in the first years of the 2000s by SWISSAID[2] and the Colombian NGO Grupo Semillas, focusing mainly on the Caribbean region of Colombia [5].

In 2002, Colombia approved the use of transgenic corn and cotton seeds, leading several Indigenous and peasant communities to declare their territories as transgenic free. For these communities and the organisations supporting them, transgenic seeds were seen as a threat to local biodiversity, people's health, and local economies [6]. The Indigenous *Resguardo*[3] of San Andrés de Sotavento was the first area to exclude transgenic seeds in 2005 and was soon followed by other *Resguardos* in the region.

Indigenous communities argued that seeds are unable to be privatised or trademarked since they really belong to mother earth [7]. They also raised concerns about the global privatisation of biodiversity, and the ongoing negotiations for a free trade agreement with the United States of America (USA) [6].

In 2009, the *Resguardo* of Cañamomo y Lomaprieta led the creation of a local seed guardianship network with the support of the Seeds of Identity Campaign. This process began by identifying local farmers that kept heirloom seeds and grouping those willing to collaborate in a network of custodians. The objective of this organisational process was to produce seeds that would enable a self-sufficient territory, capable of advancing productive projects designed by the local mayor's office and the *Resguardo.*

The Seeds of Identity Campaign supported this by strengthening the organisational effort through financing and providing technical support to improve seed quality [8]. The experience of Cañamomo y Lomaprieta served as an example of what would become a broader strategy for SGNs in Colombia. Also supported by SWISSAID, this strategy sought to identify existing native seed custodians in an area and organise them into networks or around community seed houses.

These networks would develop inventories of local seeds, characterise them to recover dwindling varieties or preserve existing ones, and then propagate and distribute them in the area. Crucially, the local networks would connect with other organisations and the Seeds of Identity Campaign to form a national seed alliance to advocate for favourable state-level policy. The alliance has, since inception: worked for the promotion of native seeds and their improvement; organised knowledge and seed exchanges at the local, regional, and national levels; developed communication strategies to emphasise the importance and potential of native seeds; and lobbied governments for the inclusion of native seeds in decision-making and public policy [9].

As SGNs were emerging in the country, the Colombian government began negotiating free trade agreements with the USA (2002) and the European Union (2007). These 'trade promotion agreements' (TPAs) were the latest step in a process of liberalisation of the Colombian economy, a process kickstarted by President Cesar Gaviria in 1990 by opening up the country to multinational corporations and removing import tariffs without protection for internal markets [10].

Particularly for food systems, this meant that large estates and monocultures were favoured while smallholdings were discouraged, accelerating the dismantling of AfroIndopeasant lifeworlds and giving way to more technoscientific knowledge safeguarded by patents [11]. More generally, this led to a 're-primarisation' of the Colombian economy (focusing especially on energy and mining), a simultaneous shrinking of agriculture and industry as part of national GDP, and an increase in unemployment and inequality [10], [12].

This process came to a head when several 'trade promotion agreements' were signed between Colombia and counterparts like the USA, Canada, and the EU in the early 2000s. In addition to preferential tariffs for food products, the free trade agreements established sanitary and phytosanitary measures, to protect human, animal, or plant life or health. Additionally, it strengthened intellectual property rights (IPR) regimes and their enforcement [13].

Of particular interest for seed systems, the TPAs required Colombia to ratify the 1991 International Convention for the Protection of New Varieties of Plants (UPOV). Up to this point, Colombia ascribed to UPOV 1978, which allowed seed saving of IPR-protected varieties by farmers. Moving to UPOV 1991 meant more protection for industrial breeder rights but impacted long-standing farming practices like seed saving and exchange [14], [15].

These requirements regarding intellectual property rights over seeds became a key factor in stimulating public debates regarding 'informal' seeds in Colombia [7]. For food sovereignty social movements in Colombia and beyond, these regulations that moved the system towards global standardisation of seeds and IPR protection for transnational corporations could 'annihilate popular food autonomy and sovereignty' [16, p. 1].

As the TPAs were being negotiated, the Instituto Colombiano de Agricultura (ICA, Colombian Institute of Agriculture) approved Resolution 970 in 2010 to establish 'requirements for the production, conditioning, import, export, storage, commercialisation, and/or use of seeds for sowing in the country and their control' [17, Sec. 1]. This piece of legislation was controversial for informal(ised) food system actors, as it was interpreted as an additional effort to privatise germplasm, effectively prohibiting peasant and Indigenous communities from saving their seeds for resowing [18]. As such, Resolution 970 was framed as a direct threat to AfroIndopeasant identities and ways of life [19], an interpretation that would massively impact the agricultural system over the next decade.

Regretfully, this period also saw Colombia facing one of its most serious environmental disasters in recent memory. Between June 2010 and May 2011, the heaviest rainfalls for over 40 years poured over most of the country. This climate phenomenon caused floods and landslides affecting 70 per cent of the national territory, damaging the homes of 2.2 million people and killing 301 people. The extent of the damage and the number of people affected led then President Juan Manuel Santos to declare an economic, social, and ecological emergency and disaster situation to deal with the impacts.

2011–2015: Fallout of TPAs, climate shocks, and novel social movements

In 2012, the trade partnership agreement between Colombia and the USA was signed, leading to the implementation of Law 1518, which binds

Colombia to the UPOV 1991 protocol [1]. In response, SGNs, which had not been consulted by the government, decided to challenge this law in Colombia's constitutional court. The courts recognised that UPOV91 could adversely affect the biodiversity, culture, and ecosystems of Indigenous and Afro-Latino communities, and sided with SGNs [7].

This was a big win for the networks, as the law was declared unenforceable. Whilst this decision did not affect other regulations that had already been introduced to implement UPOV91 (i.e., Resolution 970) [1], [20], the networks continued to work throughout 2013 when the First National Meeting of the Free Seed Network of Colombia was held [21]. The construction of the first *casa de semillas* (seed house) in Cañamomo y Lomaprieta also started that year. This first seed house was a system that included custodians, growers, and plots of land working around a 'seed centre', where seeds were selected, identified, registered, and stored to guarantee their quality for exchange, loan, or trade [8].

In mid-2013, multiple peasant organisations came together to develop a series of proposals aimed at kickstarting a national agrarian strike, a move that was directly correlated with the environmental disaster of 2010 and the signing of TPAs. The main issue for the leaders of these peasant agrarian movements was a breach of previous government commitments to food producers, including subsidies on product sales, and negotiations to discuss prices of agricultural inputs, among other issues.

Peasant organisations came together as the *Mesa Nacional Agropecuaria y Popular de Interlocución y Acuerdo* (MIA, National Agricultural and Popular Roundtable for Dialogue and Agreement). Demands of social movements included opposition to TPAs and food imports, more protections for small producers, and defence of food sovereignty and autonomy. SGNs also participated in the formulation of the national strike, producing a list of demands that included protection of farmer seeds and elimination of all IPR-related seed policies [7]. The movements consolidated the initial proposals in a document with six demands:

- To implement measures and actions responding to the agricultural production crisis.
- To guarantee access to land ownership.
- To recognise peasant territoriality.
- To facilitate the effective participation of communities and small and traditional miners in the formulation and development of mining policy.
- To comply with real guarantees for the exercise of political rights of rural populations.
- To increase investment in education, health, housing, public services, and roads for rural and urban populations [22].

The national agrarian strike started in late August of 2013 in 25 regions and with more than 50 instances of vehicular traffic interruptions. This stirred solidarity in urban areas, where other unions and social movements mobilised in support, leading to a heavy-handed response from the armed forces. After ten days, the national government negotiated with potato and dairy farmers in three regions, agreeing to not apply Resolution 970, to organise a technical dialogue about seeds, and other measures related to the TPAs [10].

It is important to highlight that peace dialogues between the Marxist FARC-EP[4] guerrillas and the government started in 2012. Throughout the agrarian strike and subsequent rounds of negotiations, the FARC-EP expressed its support for farmer organisations and criticised government responses. This was of special interest to the guerrillas, since the first item in the potential peace accord was an integrated rural reform that, among other things, sought to 'structurally transform rural areas, closing the gap between the rural and urban, and create wellness and Buen Vivir conditions for rural populations'[5] [23, p. 7].

Buen Vivir is mentioned 31 times in the text of the final peace agreement and, although not directly defined there, it can be understood in its broadest sense as post-capitalist wellbeing in harmony with nature. For Coscione and García Pinzón, the national agrarian strike was merely a symptom of a wider political transformation happening in Colombia [10]. A series of popular mobilisations, starting with an Indigenous *minga*[6] in 2008, pointed to a broader appetite for structural change.

For the authors, these popular struggles are led by the peasantry and focus on rural areas as the main scenario for confrontations between the popular sector and an oligarchic economic and political model [10]. This observation is consistent with other opinions of a 'new plural left' consolidating in the region: an open left that articulates social concerns beyond class struggles, connecting Indigenous, Afro-Latino, peasant, environmental, feminist, LGBTQ+, and other agendas with traditional Marxist concerns to advance transformation [24], [25], [26].

Due to the multidimensional crisis that occurred between 2010 and 2012, the Colombian government launched a series of strategies to reactivate the agricultural sector. Amongst the relevant agencies to implement them were the ICA, as the national agricultural health authority, and CORPOICA, a spin-off of the ICA founded in 1993 to focus solely on research. The ICA had been responsible for seed certification since the 1960s as part of green revolution strategies, including a plan started in 1983 to distribute high-quality seeds and encourage their use by small-scale farmers [27]. In the 1990s, part of CORPOICA's mission was to oversee genetic improvement programmes for new cultivars and produce seeds for these varieties. By the 2010s, seeds were framed as the primary input of food systems in government agricultural recovery strategies, leading to the development of a National Seed Plan (*Plan Nacional de Semillas*) [4].

Throughout the first years of the 2010s, seeds gained widespread attention in Colombian society. Increasing demands from industry and government soon made evident that orientation, scope, and coverage of the National Seed

Plan were insufficient. For instance, private actors requested institutional support on normative matters for seed commercialisation. Regional governments asked for research support to develop cultivars, and national-level authorities demanded official positions on the introduction of transgenic seeds to the country or on international agreements like the International Treaty on Plant Genetic Resources for Food and Agriculture (ITPGRFA). Due to these multilateral demands, CORPOICA decided to create a division exclusively dedicated to seeds, focusing on four areas of action: (i) research on seed production and genetic improvement; (ii) seed production; (iii) strategic initiatives or programmes (including the National Seed Plan); and (iv) understanding seed supply and demand dynamics [28].

While this seed-specific division was being created inside CORPOICA in 2012, informal(ised) system actors published a manifesto in adherence to a global campaign for freely accessible seeds [29]. In it, Indigenous, Afro-Colombian, peasant, and environmental organisations, and other civil society actors, condemned national seed policy and linked the food system crisis to the negotiation of TPAs.

Broadly, the manifesto argued that: (i) government policies were increasing inequality to the detriment of rural areas by fomenting big mining, energy, infrastructure, and agro-industrial projects; (ii) AfroIndopeasant practices were being criminalised by seed, IPR, and biosecurity regulations implemented in the first decade of the 2000s; (iii) these laws were unconstitutional and in violation of the rights of multiple sectors of Colombian society; (iv) corporate control over life, all forms of intellectual property rights over seeds and associated knowledge, as well as biopiracy and transgenic seeds should be rejected; (v) seeds are an essential part of life, representing the work of thousands of generations of farmers, a heritage to be transmitted to future generations that marks the foundation of Indigenous, Afro-Colombian, and peasant culture; and (vi) farmers should have the right to freely access, produce, store, exchange, and sell seeds. This last point was framed as a hedge against climate change and as a pillar of food sovereignty and autonomy.

The manifesto makes references to seed confiscations during 2010–2011, rejecting these practices by the ICA and notifying them that 'for each seed that you confiscate from us, we will make them germinate and flourish again, multiply them, spread them to walk freely with farmers through the fields of Colombia' [29, para. 9]. The document demanded public policy that favours 'living peasant seed systems' under community control, agroecology, and participatory research under peasant control. These policies would also be geared towards prohibiting privatisation and seed monopolies, as well as limiting non-reproducible seeds.

The manifesto culminated with a call to Indigenous, Afro-Colombian, and peasant communities to preserve, protect, develop, and share their seeds as a form of resistance [29]. In the weeks following the publication of the manifesto in October 2012, the first international meeting of seed guardians convened in Venezuela. A declaration was also made as part of the meeting, framing Indigenous and peasant seeds as holders of knowledge, identity, and millenary culture. The declaration condemned current treatment of local seeds, pointing to these practices as evidence of an 'insatiable, mercantilist, and ruthless economic model' [8, p. 4].

A narrative shock to the system

Some of the 2010–2011 seed confiscations and their subsequent destruction were documented by filmmaker Victoria Solano in her documentary *9.70*, which narrates the case of peasant communities whose rice seeds were confiscated by police and later burned by ICA officials. The documentary connected seed destruction to TPAs and ICA Resolution 970, producing an overarching narrative of a government protecting corporate interests by criminalising traditional seed practices.

The documentary's portrayal of the situation alarmed social movements not only in Colombia but throughout the region [30]. This situation was seen as a manifestation of their worst fears regarding absolute corporate control over seeds and the complete erosion of their means to subsist. By contrast, CORPOICA saw the documentary as outright misinformation that led to more division and mistrust within the national seed system.

The official story behind what the documentary portrayed is that the Colombian guild of rice producers asked the ICA to exercise more controls on so-called 'sack seeds'. These seeds account for around half of the 500,000 hectares of rice produced in the country but are often affected by pests or plant diseases. The rice seeds burned by the ICA in the documentary did not meet the appropriate plant health parameters and were destroyed, so the official story says, to protect other rice farmers from harm [28].

Although the burning of seeds in Solano's documentary was unrelated to ICA Resolution 970, it cemented the idea of an uncaring government legislating in cahoots with 'Big Ag' to destroy AfroIndopeasant seeds. Now with more than 1.2 million views on YouTube, the documentary has permeated the narratives of food sovereignty social movements all over Latin America.

CORPOICA admitted that Resolution 970 was concise to the point of becoming confusing, which may have led to misinterpretations that were magnified by the documentary and other actors. Resolution 970 establishes a myriad of seed types with scant definitions that end up meaning everything and nothing. Many Colombian researchers frame Resolution 970 as prohibiting native seed saving and exchange, effectively rendering AfroIndopeasant practices illegal [7], [10], [11], [20], [21], [31].

Yet despite the sheer quantity of research that supports this position, the oppositional response of social movements may have been unwarranted because, as far as CORPOICA is concerned, the resolution was never intended to regulate native seeds at all [28]. The text is incredibly ambiguous but generally points to technoscientific seed production and its use within the formal system [17].

This book aligns with positions of the Colombian state having a clear agro-industry bias, with policies skewed in its favour to the detriment of AfroIndopeasant practices. However, it also recognises that social scientists may have their own biases. Researchers that frame Resolution 970 as outlawing native seeds and related practices may be advocating in support of marginalised actors, mobilising their research to advance the AfroIndopeasant cause. While this is commendable and very much needed as a response to power asymmetries in the system, there is a clear ideological bias there that appears to completely neglect the national necessity of preserving plant health, with benefits to all farmers. While not having extensive knowledge of the more nuanced motivations of the ICA, this book recognises that these episodes may very well respond to quelling plant disease to the benefit of the broader food system.

Non-legal spaces and invisibilisation

Although not directly criminalising informal(ised) seed practices, Resolution 970 does something equally damaging: it keeps these practices in a non-legal space.[7] This piece of legislation is a good example of state actors failing to recognise non-Modern practices, a view shared by network actors like Grupo Semillas, which criticised this legislation for ignoring seeds obtained by non-conventional genetic improvement (i.e., AfroIndopeasant practices) [16].

Government invisibilisation of AfroIndopeasant food practices may be an unintended result of internalised agro-industrial biases. However, the consequences of non-recognition are still harmful. It perpetuates multiple dimensions of injustice on historically marginalised actors and stifles their possibilities for innovation together with state institutions. This limbo that traps non-Modern practices is an epistemic injustice produced by Coloniality, a system of power that sees these practices as primitive.

In 2015, CORPOICA finally replaced Resolution 970 with Resolution 3168, an update that aimed to clarify the earlier resolution's many grey areas. In this new resolution, the prohibition on saving and selling seeds only applies to certified seeds and upholds a farmer's right to save and propagate a percentage of their market-bought seeds, excluding genetically modified (GM) seeds [1].

The resolution is explicit in its exclusive application to seeds resulting from formal genetic improvement programmes, and although it also establishes general rules for seed marketing, these do not apply for native and creole varieties [28]. Despite these clarifications, some authors argue that the resolution is still overly ambiguous by referring to 'conventional' and 'non-conventional' seed production methods without specifying them [21]. Furthermore, the ICA continued with seed destruction throughout the country under Resolution 970 and Resolution 220 [31].

For social movements, Resolution 3168 is a direct result of their pressure on the government, especially from the 2013 agrarian strike and Solano's documentary [19]. This narrative of transformation of seed legislation through pressures from civil society (Indigenous and peasants especially) is also present inside AGROSAVIA [27].

Closing this half-decade, the second meeting of the Colombian free seeds network, seed guardians, and custodians occurred in mid-2015, attracting around 350 people from 150 organisations [32]. The event was framed as a space to share knowledge about native seed practices and strategise the defence of local seeds and food sovereignty. The networks organised this meeting over five stages: (i) preparatory meetings in regional nodes to discuss strategies and actions relevant to the wider mission of the network; (ii) sharing multiple local experiences of seed guardianship and agroecology while discussing the current Colombian legal regime; (iii) developing a plan for seed conservation, advocacy, and communications; (iv) field trips to nearby farms to share experiences; and finally, (v) a seed fair and barter space to interact with broader civil society.

The ultimate intention of this meeting was to strengthen weak and disconnected organisations by opening a space for knowledge exchange to consolidate the network [32]. During 2015, participatory guarantee systems were established as parallel certification mechanisms [21], leading to the publication of a methodological guide for community seed houses that contained a participatory quality guarantee system [33].

2016–2017: Peace accords, CORPOICA shifts, and novel laws

In 2016, the much-awaited peace accord was signed between the FARC-EP and the Colombian government.[8] As a direct result of this historic event, the Ministry of Agriculture developed and approved Resolution 464 in 2017, 'by which strategic guidelines for peasant, family, and community agriculture (PFCA) are adopted' [34, p. 1]. The objective of this legislation was to:

> Plan and manage integrated state action and guide social and private institutions towards strengthening the social, economic, and political capacities of families, communities, and organisations of PFCA, supported by a rural development model with a territorial focus that improves the sustainability of agricultural production and produces wellbeing and Buen Vivir of the rural population [34, p. 3].

The distinction between wellbeing and *Buen Vivir* in the text of the resolution mirrors the ambiguous language of the 2016 peace accord. Whilst this resolution also does not provide a specific definition for *Buen Vivir*, differentiating it from wellbeing is clearly important. This interpretation is reinforced by *Buen Vivir* principles that appear in the concepts and definitions section of the text [34].

First, PFCA emphasises the role of women and community, particularly Indigenous and Black communities; it also mentions a focus on family, cooperative, and community labour, and a preference for reciprocity, cooperation, and solidarity. Second, agroecology is defined as practices and a social movement focused on agri-ecosystem interactions, sustainability, and social justice, based on local/traditional and scientific knowledge interactions. Third, 'rural development with a territorial focus' and 'proximate commercialisation circuits' speak of place-based systems thinking. Fourth, 'dialogue between knowledge(s)'[9] includes local communities, farmers, and the technical/scientific community, encouraging mutual learning that produces common but non-equal understandings. Fifth, it emphasises that farmers' seeds are not subject to control or certification by the state and that they have been cared for by local and ethnic communities for generations. Finally, the legislation also defines 'participatory guarantee systems' and 'food sovereignty'. *Buen Vivir* is also hinted at in the guiding values of the resolution, which commit to 'actions and relations that respect the lives of people, nature, and knowledge' [34, p. 8].

The vision of the resolution is to position PFCA as the main driving force of rural development and a pillar for the consolidation of peace by 2038. It does so through three precedents. First, the resolution acknowledges the global and national repositioning of family agriculture and peasant economies as important for food security[10] and combating climate change. Second, it points to the importance of opening spaces for dialogue with peasant, Indigenous, Black, and other historically marginalised communities at a national level to produce these policies. Third, it criticises technical assistance initiatives as exclusionary and disconnected from the needs of communities.

The resolution also proposes ten guidelines to reposition PFCA as part of rural development: (i) rural extension and capacity building; (ii) public rural goods; (iii) access to land and land tenure; (iv) incentives and financing; (v) sustainable production systems; (vi) proximate commercialisation circuits; (vii) social marketing; (viii) non-agricultural diversification of production; (ix) advocacy and participation; and (x) information systems.

It is relevant to expand on guidelines iii, v, and vi, as they promote agroecological practices and knowledge(s), as well as dealing directly with seeds. Several proposed actions in each of these guidelines are relevant to the broader seed system transformation process. In education, a strategy for Rural Agroecological Schools will be designed and implemented; agroecological learning communities will be promoted, while technical high schools, technical and technological schools, and universities will include agroecology courses. Pedagogical tools focused on ancestral knowledge(s) and practices will be specifically developed in partnership with communities. Additionally, participatory guarantee systems will be legally recognised and regulated in collaboration between territorial agencies and communities.

Regarding seeds, the resolution recognises the loss of native and creole seeds, as well as a reduction in the capacity of communities to handle them autonomously. The resolution aims to 'establish and support networks that preserve, guard, defend, and reproduce farmers' seeds as a measure to resist the effects of climate change, and plant diseases, preserve agrobiodiversity, traditional practices, and knowledge(s) of PFCA' [34, p. 12]. Other relevant actions proposed in the resolution are: a strengthening of community seed houses through participatory mechanisms and funding, so that PFCA actors can access subsidised or free resources; promotion of participatory research in the recovery, conservation, and use of farmers' seeds; and support of transgenic-free areas while strengthening controls over corporate appropriation of farmers' seeds.

Emergence of AGROSAVIA

In parallel with the publication of Resolution 464 by the Ministry of Agriculture, a series of broad internal transformations occurred in CORPOICA, starting in 2017. The appointment of a new general director led to the development of a new corporate mission, vision, and objectives. Research focus and practices also began to shift towards a more integrative, holistic, and participatory approach. CORPOICA started hiring more anthropologists and sociologists, moving from a highly technical focus to include the social dimension of farmers' lives. This opened more space for participatory action research to validate technologies provided to farmers by CORPOICA.

The organisation also grew more preoccupied with expanding its involvement in the value chain to include food consumers, understand food markets, and consider how to create new market opportunities for farmers. This meant moving from producing demand-driven technologies to creating the demand for fresh and processed food products. A culture of collaboration was also encouraged, both internally and externally, manifesting in researchers seeking partnerships and joint funding opportunities with other organisations, more interdepartmental dialogues, and framing projects as farmer-researcher partnerships instead of top-down impositions [35].

Despite these internal changes, CORPOICA still faced the enormous challenge of overcoming widespread mistrust. Historically, the organisation had a poor reputation amongst farmers throughout the nation and across different crop types [28], [35]. This issue was directly related to CORPOICA's heavy top-down approach, with researchers directing projects without, it would seem, much connection to the lived experience of farmers [35]. Moreover, narratives emerging from historical institutional distrust continued to erode collaborative possibilities, e.g., AfroIndopeasant seed criminalisation and other issues as portrayed in the documentary *9.70*.

Many social movements in Colombia also believe that CORPOICA willingly surrendered national seed biodiversity to multinationals. This narrative stems from farmers who requested access to seeds from the national germplasm banks and had their requests denied. As clarified by CORPOICA, the role of the germplasm banks is to store low quantities of seeds (often just 100 grams or 100 seeds of each variety) for seed production and improvement programmes. The reality is that anyone can access these seeds, but they first must comply with specific viability and conservation standards [28]. The cumulative image problem built on these narratives ultimately led to a rebranding of CORPOICA as AGROSAVIA in 2018, marking the 25th anniversary of the organisation.

Within AGROSAVIA, seed-related work had been very much the initiative of individual researchers without an organisational structure. In 2017, the Ministry of Agriculture framed seeds as being of strategic importance for the country and meriting being positioned as a vital line of work. Accordingly, it published a regulation mandating the creation of a specific group working on seeds. A new seed department was formed in AGROSAVIA involving staff from all regional centres.

The seed department was created as a complement to other teams working on seeds, e.g., seed breeding and conservation programmes led by the agro-biodiversity group, and laboratory-based seed production and rescue overseen by the laboratory department. Articulation between different areas was a priority for the seed department to complement their work with diverse expertise. The culture emphasised within the department was that their work retained relevance at a national level, adopting a reflective stance regarding their impact on the daily lives of farmers. This new area in AGROSAVIA appeared to have a thoughtful approach to farmers' lived experience, translating into a heavy emphasis on responsibility and accountability for how their decisions may affect multiple rural actors [28].

To lead the new seed department, AGROSAVIA assigned a person with expertise in rural development with a wider perspective on seeds to encourage such an integrative approach. The role was expected to connect technoscientific knowledge of seeds with strategic thinking, and the daily needs of farmers. Expectations also included the encouragement of dialogues between diverse actors in various scenarios by understanding the multiple dimensions of seeds (technology, uses, culture, identity, knowledge, etc.) [28].

A two-pronged strategy: Seed plan and system

AGROSAVIA's seed department currently works on two fronts: (i) the National Seed Plan and (ii) the National Seed System. The National Seed Plan is an internal macro-project that seeks to build seed production capacities in farmer organisations. AGROSAVIA provides knowledge to strengthen farmers as quality seed producers, including genetic improvement training. The agency prioritises crops like rice, corn, beans, sugarcane, potato, *chontaduro* (peach palm), and avocado, among others. The National Seed Plan currently involves 60 organisations in 19 regions, and aims not only to fulfil local demand but to cover other national and international markets.

AGROSAVIA feels that some local organisations of seed producers saw the National Seed Plan not only as a path towards food security but also as a viable source of income. Some saw it as an opportunity to revitalise species like the *pajarito* pea of the Colombian Atlantic coast. For others, it was the first time they were approached by a government agency to support local agriculture, as some communities are extremely remote, only reachable after hours on muleback from the nearest town [28]. However, not all communities were equally supportive of AGROSAVIA's seed plan. When it approached them asking for seeds with a promise of improving and returning them with full certification, many communities saw this as seed appropriation and a threat to seed sovereignty [28].

As SGNs started hearing of AGROSAVIA's approaches, they invited the organisation to present the National Seed Plan to a forum organised by the Alliance for Biodiversity. This is a multi-actor space where NGOs, academics, universities, social organisations, SGNs, and all those interested in defending native and creole seeds gather to discuss relevant issues.

This meeting led the Organización Nacional Indígena de Colombia (ONIC) to approach AGROSAVIA to develop a project on Indigenous seeds supported by the Ministry of Agriculture. However, due to a lack of organisational experience in agroecology, the project was not initiated. This situation has changed in the last years, as AGROSAVIA is involving more experts on agroecology [36]. Parallel to this initiative, SGNs proposed a similar project focused on agroecological seeds.

The 'Colombian seed system' (*sistema de semillas en Colombia*, also known as the 'national seed system') is a much broader concept that includes all preexisting activities and relationships between its actors. For AGROSAVIA, the objective of the Colombian seed system should be to increase the availability and use of quality seeds, benefiting from their potential while reducing risks from poor-quality material [28].

The organisation believes the fulfilment of this objective requires recognition of the widely diverse food production systems that exist in Colombia. This stems in part from questioning the capacity of the formal seed system to attend to all Colombian seed production [27]. AGROSAVIA acknowledges that all agricultural practices and their preferred seed types (native, creole, regional, improved, transgenic, etc.) are relevant and important.

To dynamise the Colombian seed system, AGROSAVIA deployed a two-pronged strategy. On one side, the National Seed Plan strengthens local systems with organisations producing quality seeds. On the other, a top-down effort aims to bring system actors together, aligning them with the concept of a national seed system, and collaborating towards its development. The organisation believes the Colombian seed system is beginning to consolidate and hopes to see a system where actors recognise each other and coexist harmoniously. The organisation has kept working to eliminate polarisation in the sector between the discourses of improved seeds produced by multinational corporations and the creole/native seeds of social movements [28].

For AGROSAVIA, all seed actors should recognise the need for research, development, education, and normativity beyond their preferences or particular seed inputs [28]. This strategy is also supported by a study of national seed systems in other countries done in collaboration with the Food and Agriculture Organization of the United Nations (FAO-UN), and a retrospective/prospective study of the Colombian national seed system.

2018–2022: Co-visioning a national seed system

As the seed department was consolidating, AGROSAVIA started to pitch the idea of a national seed system to other relevant organisations [28]. Complexity and fracturing in the system[11] led AGROSAVIA to favour one-on-one detailed conversations instead of opening many fronts of action to advance the process. This meant the department opted against an open call via mass emails.

The enlistment process was slow, as AGROSAVIA carefully told the story of an emergent national seed system to each actor one by one. The start of COVID-19 restrictions also contributed to a sluggish recruitment process, forcing dialogues into a virtual space. Rather than moving slowly, one could say that the process moved at the speed of trust as discussions were laborious and extremely complex even with single actors [28].

To efficiently advance this process, the organisation prioritised actors who would most probably have a positive response. The initial group included the Ministry of Agriculture, FAO-UN, and ACOSEMILLAS (the Colombian association of seeds and biotechnology). With their involvement, AGROSAVIA started contacts with ONIC and SWISSAID, organiser of the Campaña Semillas de Identidad and promoter of SGNs in Colombia. Involvement of these actors required a gentle touch by AGROSAVIA, due to decades of mistrust, recognising points of agreement while also establishing clear demarcation lines.

This process is ongoing and AGROSAVIA expects to enlist other actors that are perceived as more difficult at a later time [28]. This overly cautious approach is understandable, but may not be fully justified. SGNs appreciate that AGROSAVIA is using a complex systems approach and making an effort to map all actors involved in seed processes. What SGNs really expect is a vision built in a participatory manner without dominance from government or business actors [36].

The seed department has also encouraged interdepartmental collaborations within AGROSAVIA to interact with external actors. For example, when interacting with ONIC, the department assembled a team including legal, intellectual property, technical, and agronomy experts. This integrated approach is seen as a winning strategy by the department, allowing them to speak in diverse contexts, and to move from claiming absolute truths to being supported by experts in specific areas relevant to seeds.

The seed department also decided to promote evidence-based discussions by sharing project results from the National Seed Plan [28]. This proves that AGROSAVIA is involved throughout the seed system (e.g., rescuing creole varieties but also doing genetic improvement – including transgenics, or developing agroecological production systems but also using chemical fertilisers).

There has also been a clear effort by the seed department to be transparent. All actors involved are aware of each other's participation and AGROSAVIA constantly emphasises that the country needs all actors and seed types. Inclusion of all participating actors will require fine-tuning institutions and norms, but consensus should occur within a common seed system framework [28].

Co-visioning workshop process

To advance this co-visioning process of a national seed system, AGROSAVIA held a first workshop with 16 actors in late 2021. Due to COVID-19 restrictions, this encounter was a hybrid event, held online and in person, with participants from government, industry, guilds, and non-profits. The workshop began with AGROSAVIA representatives sharing their ideas and intentions, emphasising that they do not want to impose an ideal seed system, but seek to create the system Colombia needs [28]. Participants were then organised into working groups that paired antagonistic organisations to gather data on their visions, previous approaches, and proposals for next steps.

This workshop was far from easy for AGROSAVIA, as some actors held preconceived notions that other participants had hidden agendas. The seed department had to emphasise the common interests over seeds and the previous socialisation phase with participants. In addition to these difficulties, the department was still struggling with virtual tools and workflows [28].

Despite these obstacles, the workshop allowed points of agreement between system actors to surface. All believe that seeds are an important national resource, especially since Colombia is one of the most biodiverse countries on the planet; all actors understand the natural richness embedded in seed agrobiodiversity. Seeds are also seen as deeply connected to food security and cultural identity. Actors also agree on quality as an integral part of seed production. However, there are disagreements regarding existing norms, legislation, and their implementation [28], [36].

Remaining tensions in geography and law

One contentious aspect of ongoing dialogues concerns the interaction between transgenic and non-transgenic seeds. One of the key issues for SGNs is to maintain some territories free of transgenics, a measure that would allow the creation of distinct areas for agroecological and organic farming [36]. These types of agriculture become impossible when there is a risk of contact with agrichemicals or transgenics. The networks use the example of big landowners using airborne fumigation of their crops, which often affect the health and production of surrounding small farmers.

For the networks, public policy regarding land use should be developed to address this issue in seed systems [36]. AGROSAVIA does not see an easy way to resolve this tension, as their programmes and technicians should be able to reach any region in the country [28]. Additionally, this could limit free circulation of hybrid and transgenic seeds, or essential inputs like fertiliser, ultimately leading to individual conflicts between neighbours that choose to grow food in different ways.

Coexistence is important for the networks, but they are well aware of power asymmetries in the system, which is why they are pushing for novel policy frameworks that protect their practices [36]. However, this position by the networks may prove problematic in the co-visioning process. AGROSAVIA feels it has been able to sway organisations like the ICA and ACOSEMILLAS to recognise the importance of creole and native seeds. The seed department sees a favourable discourse emerging in industry actors of coexistence with 'informal' seeds, opening spaces for novel certification schemes and other inclusive regulations aligned to the realities of these seed types [28].

A push for land reform limiting seed use may be a step too far for industry and government. Despite these unresolved tensions, AGROSAVIA is optimistic about the process, since a wide range of actors are now talking about the emergent system, and those currently involved are open to dialogue [28].

Differing measures of seed quality

Seed diversity is important since Colombia has a wide variety of ecosystems and soils. Seeds are adapted to specific conditions, so a single seed that will work everywhere is impossible. Yet there is a simplistic view in the system whereby industrial seeds are framed as being inherently good and native seeds inherently bad, with some even tying this to racism: 'Because these [industrial seeds] are produced by whites, and these [creole seeds] are produced by blacks' [36].

Seed networks emphasise that although AfroIndopeasant actors sometimes use low-quality seeds carrying pests or diseases, or that produce lower yields, these creole seeds could also be considered high-quality seeds. An essential part of the co-visioning process, therefore, is to achieve consensus on what constitutes a 'good' or 'bad' seed, inclusive of the huge diversity of native and creole seeds [36].

Qualifications of good or bad seeds must acknowledge existing agro-industrial biases and state–industry interactions that have shaped the existing seed system to benefit some more than others. As Silva Garzon demonstrated in the case of transgenic cotton, GM seeds are not a panacea and may be sometimes detrimental to Colombia's seed system [37]. Incumbents typically frame the value of native and creole seeds as purely symbolic or as part of national genetic resources.

This tends to obscure the fact that these seeds guarantee food sovereignty, security, and income as a fundamental part of the economy of many families. If those seeds were of bad quality, the survival and livelihoods of millions of people would simply not be possible. As mentioned by Gutiérrez Escobar and Fitting, these narratives challenge hegemonic discourses of local seeds as 'raw material and a resource, to be "discovered", "invented" and commodified by industry and Western-based science' [7, p. 712].

For SGNs, appropriate quality metrics should be developed for these seeds, distinct from those used by industry. One important aspect would be to decouple yield volumes from the inherent quality of a seed [36]. For example, hybrid corn seeds may produce yields of up to 18 tons per hectare, while creole varieties may yield one or two tons. For the networks, this should not be taken as a sign of low quality. The seed department at AGROSAVIA agrees that specific seeds produce particular yields and should be studied to develop standards [28], [36].

Another important aspect to consider in alternative quality metrics is the varied uses of crops in AfroIndopeasant agriculture. In these food systems, there is no need to produce a standardised product aligned to market demands. Using corn as an example again, all of the crop would be put to use depending on its characteristics: some would be moved to market, some would be used for home cooking, some to feed poultry, etc. [36]. In these settings, quality takes on a different meaning since homogeneity of the crop is not very relevant.

Narrative battles

At the intersection of seed quality and transgenics, it is relevant to review the discursive battles that circulate alongside them. An outstanding example is Silva Garzon's study on the introduction of transgenic cotton to Colombia. The author maps three narrative strategies used to monopolise transgenic seed markets: choice, care, and discredit[12] [37]. 'Choice' assumes cotton farmers are able to choose transgenic seeds freely whilst, in reality, their expansion was accompanied by seed confiscations that led national seed producers to temporarily exit the market. 'Care' frames the farmer as the leading force in crop development; if seeds are successful, it is because they are good and stable, but if the crop fails, it is because of undisciplined farmers. Extreme weather is also used as an explanation when farmers do not get the expected yields from transgenics.

It is common for technicians to use the expression *mal del trópico* (ailment of the tropics) to refer to foreign seeds' lack of resilience in rough local ecosystems [37, p. 13]. As Silva Garzon suggests, while the ailment of the tropics highlights the fact that not all seeds respond in the same way to the same ecosystems and climates, it is also a 'miasmatic argument with colonial undertones that describes the tropics as wild and untamed' [37, p. 13]. Finally, the narrative of 'discredit' frames foreign seeds as vulnerable and local seeds as robust, emphasising the roughness of the place where they are sown instead of their inherent quality.

This adds to existing framings of 'creolised seeds'[13] [37, p. 14] as 'conventional' while transgenics are sold as 'unique' seeds evoking processes of technological innovation [37]. These narratives dominate the repertoire of monopolisation processes and are also effectively deployed to assign responsibility for seed success and failure throughout the food system. The narratives unveiled by Silva Garzon connect to the deeply ingrained agro-industrial and technoscientific bias in Colombian institutions, biases that need to be transcended in the co-creation of a national seed system.

Epistemic injustice

Knowledge is key in the seed system transformation process, and SGNs have also noted agro-industrial and technoscientific biases in this space. This suggest there is an overarching issue of epistemic injustice in the system. AGROSAVIA places a high value on scientific knowledge and technical results but also recognises that the scientific method is not the only form of knowledge that can support recommendations.

The seed department recognises, respects, and values ancestral knowledge, but feels this knowledge is often not made sufficiently explicit by communities, effectively hindering development processes [28]. For their part, SGNs advocate for the development of a 'real knowledge' of seeds, where each of their characteristics is known. This would facilitate evaluations of their multidimensional quality instead of using broad generalising value judgements of seeds as good or bad [36].

Debates regarding seed knowledge must also consider the contextual variables of a country like Colombia. In the case of small-scale farmers, knowledge comes from their parents, grandparents, or neighbours and is transmitted from generation to generation. These farmers do interact with technicians, but most knowledge transfers occur from local knowledge-holders who have experience with specific crops. For big landowners with highly industrialised crops, knowledge may come exclusively from technicians, mainly agronomists who manage production systems, while landowners focus on commercialisation. Adding to this complexity, agronomists usually work for local commercial distributors selling chemical inputs that are the product of technoscientific institutions aligned with the dominant model of agriculture (green revolution, monoculture, etc.) [35].

AfroIndopeasant, agroecological, and organic agricultural knowledge is still very much alternative, marginal even. Vested organisations have very little access to resources for research, development, and knowledge transfer [36]. This power asymmetry is also evident in the advocacy work done by SGNs when interacting with government. The networks are constantly pushed to demonstrate that agroecology or native/creole seeds are good and have positive impacts in the economy, the environment, and people's health [19], [36].

For some network actors, demands to prove these practices are useful or good for the environment are an absurdity, especially when the requirements for big companies regarding new transgenic technologies or agrichemical impacts appear more lax. The networks think it is a travesty that powerful actors require them to carry the burden of proof, having to constantly produce evidence persuading others that they are not harming people or the environment. This continues even when native seeds and AfroIndopeasant or agroecological practices have been feeding communities for decades, centuries, or millennia.

This situation is slowly changing, as actors like FAO-UN do more work to research and promote agroecology. SGNs themselves are also compiling cases to make benefits visible, often in collaboration with external actors that value these practices (e.g., chefs, or health and nutrition advocates) [36].

Carrying the burden of proof to demonstrate the value of AfroIndopeasant agricultural practices also constitutes an epistemic injustice (see Chapter 9). Existing biases in the system lead to an impairment of AfroIndopeasant knowledge achievements. Additionally, these actors do not have the resources to communicate this information widely, resulting in multiple harms perpetuated by the system [38].

The differences in knowledge practices and resulting power imbalances discussed in this section may prove difficult to navigate in the co-visioning process but are of the utmost relevance to the successful transformation of Colombia's seed system. For Vidal and Escobar, the aforementioned conflicts over seeds in Colombia could be framed as processes of cultural Coloniality, defined as 'a form of epistemic-political domination related to the issue of expertise and knowledge inclusion/exclusion processes in the production of public policy' [20, p. 44].

There are two key premises here: (i) there is a single knowledge that can be legitimised when developing seed public policy; and (ii) this knowledge can only be recognised economically when it is found as private property (in this case, defined as intellectual property) [20]. This is also evident in Silva Garzon's study on cotton, where narratives are often deployed that lay the blame for GM crop failure and seed effectiveness on farmers or environmental conditions, while transgenic seeds are given a pass since they are framed as inventions unable to fail [37].

The complexity of the co-visioning process cannot be overstated; transcending mistrust in the system and pre-existing conceptions of other actors' agendas is a monumental challenge. However, despite the deep differences between perspectives, there seems to be an overall agreement on coexistence emerging. All actors involved see each other's practices, narratives, and materials as valuable to food provisioning in Colombia. They recognise the multiple dimensions of seeds, their importance, and the need for quality seeds throughout. Disagreements appear to orbit around rules: these might be differences in interpreting existing legislation, implementation of the emerging seed system through laws, who should govern it, and norms on transgenics and transgenic-free zones.

Knowledge is an important part of the process, but navigating it in its plurality is challenging. There are still latent colonial power structures that perpetuate epistemic injustices, such as the way in which technoscientific knowledge is represented as dominant. This narrative manifests in creole seeds being perceived as low-quality despite being an integral part of PFCA systems, as well as in burdening AfroIndopeasant communities with proving the value of their knowledge(s) and resulting technologies, despite millennia of successes. Despite persistent Modern/Colonial biases, there seems to be one point of agreement: incumbents want non-hegemonic knowledge to be more explicit, while informal(ised) system actors want to co-produce better knowledge of native seeds in their wide diversity.

2022–: National seed system futures

The continuing co-creation of a national seed system coincided with a radical shift in the national government, the election of the first leftist president in the country's history. The political project of the 2022–2026 government was informed by *Buen Vivir* principles and sought to raise the voices of historically marginalised sectors. However, the implementation of their agendas has been lacklustre. Regarding the national seed system, tensions between worldviews remain and negotiations are bogged down by bureaucracy, populism, and erratic decision-making [39]. For instance, policy debates might provide only two-minute slots for actors to communicate their positions. Leftist emancipatory impulses might neglect important technical aspects of seeds. And, crucially, there have been unfulfilled promises regarding government funding for seed houses. A salient example of difficulties in decision-making is that the position of directorate of innovation at the Ministry of Agriculture has changed ten times in three years [39].

The arrival of Buen Vivir

As defined in Chapter 2, and expanded in the next two chapters, the narrative of *Buen Vivir* has been extremely influential in policy change in the Andes. Although Colombia was late to the trend, having a leftist government in power signalled the arrival of a *Buen Vivir* politics. In their victory speeches, the 2022–2026 president and vice-president made several mentions of *Vivir Sabroso*, which roughly translates as 'living flavourfully'.[14] This phrase can be interpreted as a play on *Vivir Bien* (roughly, 'living well'),[15] adjusted to the culture of the Colombian coasts from which the candidates originate.

The speeches emphasised the role of marginalised racial and ethnic identities and ways of life, LGBTQ+ identities, feminism, the importance of water and, crucially, the rights of nature. However, President Gustavo Petro appears to align more with a classic left focus of industrialisation of agriculture, even mentioning that he wants to develop Colombian capitalism. Government narratives appear to suggest intentions to integrate Modern and non-Modern perspectives. However, implementation of *Vivir Sabroso* institutions has been lacking as the tenure of this government is ending.

Strengthening and legitimising

In the long term, AGROSAVIA sees a mature national seed system consolidating, and hopes that strengthening it will lead people at higher decision-making levels to better understand the value of seed supply. The seed department wants people throughout the country to understand the importance of seeds and their quality. In their opinion, people are too focused on normative issues and should be more attentive to seed benefits and risks [28].

For this to happen, the concept of a national seed system needs to be legitimised. For the networks, legitimacy will come through visibility, and they feel AGROSAVIA has been working towards this through the co-visioning meetings. Some feel the outputs of these meetings should be published as an official document, enshrining the system in public policy that recognises, protects, and stimulates AfroIndopeasant practices [36].[16] Yet the networks fear that this will become just another academic exercise without much impact on the system.

For SGN actors, true legitimacy would come through development of a new public policy that promotes coexistence in the system while supporting native seeds and the networks. AGROSAVIA knows that the seed system will eventually permeate legislation but is not convinced that a specific law would be the best way to implement it. Such a law may be perceived as yet another imposition rehashing the tensions of previous normative attempts. However, steps are being taken to produce a legal framework that opens more spaces for alternative practices.

In 2023, the Constitutional Court published Sentence T-247 affirming that the Colombian state had failed in protecting the biocultural patrimony of native and creole corn seeds [40]. This Sentence was the product of Indigenous peoples reclaiming their rights to food sovereignty, particularly to protect local corn varieties from transgenics. Crucially, it led to recognition by the state that there is no public policy to recognise and protect native seeds, as well as no specific regulation for their production and commercialisation. Sentence T-247 pushed the Ministry of Agriculture and related agencies to create a legal framework for these purposes. This clashed somewhat with the co-creation process already underway in AGROSAVIA as it changed the conditions and actors involved [39].

In 2024, a guiding document on agroecological agriculture was published by the Colombian government [41]. The document was developed with the participation of AfroIndopeasant and agroecological organisations, as well as researchers. It presented the technical foundations, orientations, and strategic objectives to formulate public policy for agroecological agriculture. The document frames agroecology as a systemic transformation including cultural, economic, political, and environmental dimensions. Crucially, it recognises the lack of data on agroecological production (i.e., the invisibilisation of existing actors, value chains, territories, etc.).

The guiding document aims to: (i) strengthen agroecological production and those involved in it, (ii) ensure cultural and environmental sustainability, (iii) reduce inequalities in access to resources, and (iv) transform the agri-food system towards local, healthy, and sustainable models. To achieve this, it proposes integration between peasant and scientific knowledge, a transition towards more diversified production, strengthening alternative markets (short, local, solidary), and creating the institutional and normative conditions to implement this agroecological shift [41].

Sentence T-247 and the guiding document on agroecological agriculture set the stage for a more Transmodern seed system, particularly expanding participation of historically marginalised actors in its governance. Specific policies and regulations are yet to emerge and challenges abound, including defining common terms (e.g., what is a native or ancestral seed), achieving coexistence of diverse actors with contradictory worldviews, and defining seed quality [39]. Another important aspect for the future of the system is the question of ownership. AGROSAVIA sees itself as just another actor, rather than the de facto administrator of the national seed system. The organisation would most likely become the main body to develop a system proposal, but all participants should agree on who would run it; for the seed department, the natural fit for this task is the Ministry of Agriculture, as AGROSAVIA cannot be left as the judge, jury, and executioner [28].

Continuing the seed plan

Seeds continue to be a priority for AGROSAVIA, and their seed plan is growing. The plan is reaching more than 100 communities in 22 states working with agroecological nodes (including SGNs), peasant and Indigenous communities. The plan has led to the creation of intercultural seed production protocols integrating AfroIndopeasant and scientific knowledge. Additionally, AGROSAVIA has engaged in acts of seed rematriation, moving germplasm from the national seed bank to territories that have lost agrobiodiversity due to environmental catastrophes or displacement caused by the armed conflict [39], [42]. Salient examples include the sowing of 165,000 native seedlings in the Atrato River (which gained recognition as a subject of rights in 2016), the recovery of native cotton varieties by Indigenous communities of the Sierra Nevada, and rematriation of corn seeds to Kamëntsá peoples [39], [43], [44].

Communication and funding problems

Beyond the agreements that dialogue between AGROSAVIA and SGNs have produced, the networks visualise a difficult road ahead to specification or legislation of rules [36], as they are cynical about AGROSAVIA's motives.

Despite their cynicism, there is evidence of AGROSAVIA's intention to dynamise the system while allowing coexistence of its formal and informal sub-systems [27]. As a move towards greater understanding and agreement, asynchronous dialogue between AGROSAVIA and the networks is happening, whereby each questions the other to gauge positions and establish points of agreement and disagreement. This dialogue includes topics like transgenics, seed quality, certification, and participatory guarantee systems, among other issues [28], [36].

AGROSAVIA has also signed a cooperation agreement with SWISSAID that would allow both organisations to collaborate on future projects. However, this agreement has not been very fruitful due to funding issues,[17] which may constitute a serious obstacle for the overall visioning process. Internally, AGROSAVIA's limited resources often lead to researchers contributing money from their own pockets to pay for participant transportation or catering. The organisation has called on its researchers to always seek external sources of funding for their projects, especially from international cooperation agencies. For their part, farmers often do not have enough resources to participate in these processes, since even a small expense on transport or internet credit may be too much to bear [35].

Conclusion

So, what has contributed to change? (see Figure 4.1) Food provisioning in Colombia is split in half between AfroIndopeasant, family, and community agriculture on one side and industrial production on the other. The intensification of neoliberalism in the country led authorities to implement regulation in line with international seed rules. As a response, seed sovereignty social movements emerged to protect and strengthen existing informal(ised) micro-institutions. Due to unclear seed regulation, a narrative of AfroIndopeasant criminalisation consolidated throughout the country, emboldening social movements. This narrative painted a picture of an uncaring government allied with Big Ag to erode AfroIndopeasant practices and outlaw native seeds. This was a turning point that strengthened SGNs and their alternative practices aimed at circumventing formal institutions.

As this was happening, the peace accord between the FARC-EP and the Colombian Government deployed *Buen Vivir* into national politics. The peace accord effectively opened new political possibilities to allow for other discourses and novel policies. As a case in point, Resolution 464 backing peasant, family, and community agriculture was informed by the accord and *Buen Vivir* principles.

Finally, an internal cultural transformation in AGROSAVIA led to more inclusivity and dialogue between hegemonic and non-hegemonic sides of the seed system. The simultaneous strengthening of SGNs and the social turn inside AGROSAVIA opened a translational contact zone where Modern and non-Modern perspectives engage in dialogue and agonism to co-produce a new seed system. With the implementation of policy guidelines for agroecological food production, these novel inter-epistemic spaces might clear the way for Transmodern futures in the Colombian food system and beyond.

Notes

1 Colombian Corporation for Agricultural Research.
2 SWISSAID is a non-profit based in Switzerland that develops projects with smallholder farmers in developing countries.
3 *Resguardos* are territories that are tied to Indigenous communities and are ruled by special autonomous statutes.

4 FARC-EP stands for Fuerzas Armadas Revolucionarias de Colombia – Ejército del Pueblo (Revolutionary Armed Forces of Colombia – People's Army).
5 This and all subsequent translations from the Spanish are by the author.
6 *Minga* is an ancestral Indigenous institution used to gather the community around a common purpose, be it gaining consensus or working collaboratively on building infrastructure.
7 At the moment of writing, the ICA has not yet produced any sort of regulation pertaining to these seed types. For the time being, they remain in a non-legal space.
8 These accords put an end to a conflict with more than seven million victims and an estimated 220,000 deaths. At the time, the FARC-EP guerrillas were the longest-lived active rebel group in the world.
9 The Spanish phrase is *diálogo de saberes. Saberes* is the plural form of *saber* (knowledge). Since there is no plural word for 'knowledge' in English, the (s) is added.
10 Resolution 464 reiterates the importance of AfroIndopeasant agricultural practices by recognising that they produce more than 70 per cent of the country's food. This number is contentious, since other accounts estimate that nearly half of the food consumed in Colombia is imported [11]. However, in all accounts PFCA, Afro-Colombian, and Indigenous practices seem to account for around half of the country's food production.
11 For instance, the networks have seen businesses and government historically degrade native and creole seeds by equating them to low quality and low productivity [36]. This negative image is the opposite of what the networks seek, which is to value these seeds [19], [36]. Another example would be the dialogue with the Universidad Nacional de Colombia (National University of Colombia), where anti- and pro-transgenic camps are often at odds [28].
12 Author's translation from the Spanish terms *elección, cuidado*, and *descalificación.*
13 Silva Garzon uses the term *semillas acriolladas* (creolised seeds) to refer to seeds resulting from the adaptation of certified or industrial seeds to local environmental conditions.
14 The term *Vivir Sabroso* has been used widely in Black resistance movements of the Colombian Pacific coast and the Atrato River [45].
15 The term *Vivir Bien* is used mainly in Bolivia as a stand-in for *Buen Vivir* in order to draw a distinction with the Ecuadorian political project of that name.
16 Publication of the outputs of the co-visioning meetings might also alleviate the confusion of some seed guardians, who were baffled at the start of the process since they had never encountered a document that talked about a national seed system [19].
17 SWISSAID is ready to contribute money, but is not willing to fund projects unilaterally, while AGROSAVIA has struggled to secure resources to advance common interests. SWISSAID did fund a trust-building space as part of the national seed system co-visioning between AGROSAVIA and seed networks, but at the moment that is the only result from the agreement [36].

References

[1] D. Silva Garzón and L. Gutiérrez Escobar, 'Revolturas: Resisting multinational seed corporations and legal seed regimes through seed-saving practices and activism in Colombia', *Journal of Peasant Studies*, vol. 47, no. 4, pp. 674–699, 2020, doi:10.1080/03066150.2019.1668780.

[2] D. Rodríguez Goyes, '"Tactics rebounding" in the Colombian defence of seed freedom', *International Journal for Crime, Justice and Social Democracy*, vol. 7, no. 1, pp. 91–107, 2018, doi:10.5204/ijcjsd.v7i1.425.

[3] V. Solano, Dir. *9.70*, 2013. [Film].
[4] AGROSAVIA, ICA, and SENA, 'PLAN NACIONAL DE SEMILLAS (PNS) Programa de fortalecimiento al sistema nacional de semillas producto de la intervención de sistemas locales de semillas', 2019.
[5] X. Chacón, M. García, and Redes de custodios y guardianes de semillas y casas comunitarias de semillas nativas y criollas, 'Guía metodológica', Bogota, Colombia, SWISSAID y Corporación Biocomercio Sostenible (in Spanish), 2017. [Online]. Available: www.swissaid.org.co/wp-content/uploads/2022/05/guia-redes-de-semillas-2017-1-.pdf.
[6] SWISSAID Colombia, 'Zonas y Territorios Libres de Transgénicos', 2012. [Online]. Available: www.swissaid.org.co/wp-content/uploads/2022/05/Guia_declarar_territoriors_libres_transgenicoss.pdf.
[7] L. Gutiérrez Escobar and E. Fitting, 'The red de semillas libres: Contesting biohegemony in Colombia', *Journal of Agrarian Change*, vol. 16, no. 4, pp. 711–719, 2016, doi:10.1111/joac.12161.
[8] GRAIN, 'De guardianes, ferias y casas de semillas', 2017. Accessed: 12 Jun. 2022. [Online]. Available: https://www.grain.org/media/W1siZiIsIjIwMTcvMDgvMDgvMTBfMzZfMzNfOTY2XzAyRGVfdW5fdmlzdGF6b195X211Y2hhc19hcmlzdGFzLnBkZiJdXQ.
[9] SWISSAID Colombia, 'Semillas criollas, alimentando la vida', 2019. Accessed: 20 Jun. 2022. [Online]. Available: https://www.semillasdeidentidad.org/es/publicaciones/semillas-criollas-alimentando-la-vida.
[10] M. Coscione and V. García Pinzón, 'TLCs, paro nacional agrario y movimiento social en Colombia', *Nuevo mundo mundos nuevos*, Sep.2014, doi:10.4000/nuevomundo.67127.
[11] L. Santana Garzón and M. González, 'Rentabilidad o respeto por la tierra? Cronología de la empresarización y despojo de semillas en Colombia', 2013. [Online]. Available: www.etnoterritorios.org/apc-aa-files/d2189931b0f3b7fa37ab663a062fb165/rentabilidad-o-respeto-por-la-tierra.docx.pdf.
[12] M. Drago, 'El cerco legislativo y normativo de los tratados en América Latina', *Biodiversidad Sustento y Culturas*, vol. 93, pp. 10–13, 2017.
[13] Office of the United States Trade Representative, 'Trade Promotion Agreement, United States–Colombia', n.d. Available: https://ustr.gov/trade-agreements/free-trade-agreements/colombia-tpa/final-text.
[14] E. Senini, 'Farm saved seeds: A right to use or a right abused?' *European Food and Feed Law Review*, vol. 13, no. 2, pp. 116–124, 2018.
[15] T. Wattnem, 'Seed laws, certification and standardization: Outlawing informal seed systems in the Global South', *Journal of Peasant Studies*, vol. 43, no. 4, pp. 850–867, 2016, doi:10.1080/03066150.2015.1130702.
[16] Grupo Semillas, 'Las leyes de semillas, aniquilan la soberanía y autonomía alimentaria de los pueblos', 2010. Accessed: 20 Jun. 2022. [Online]. Available: www.semillas.org.co/portal/leyes/Internacionales/14.%20grain-4098-las-leyes-de-semillas-aniquilan-la-soberania-y-autonomia-alimentaria-de-los-pueblos%20(1).pdf.
[17] ICA, *Resolución 970*, 2010. Accessed: 14 Jun. 2022. [Online]. Available: https://www.ica.gov.co/normatividad/normas-ica/resoluciones-oficinas-nacionales/resoluciones-derogadas/resol-970-de-2010.aspx.
[18] NET002, 'private communication', Oct.2022.
[19] NET003, 'private communication', Mar.2021.
[20] N. H. Vidal and L. G. Escobar, 'Epistemic and political struggles against the privatization of seeds and collective knowledges', *Revista Colombiana de Antropologia*, vol. 55, no. 2, pp. 39–63, 2019, doi:10.22380/2539472X.798.

[21] V. García López, O. F. Giraldo, H. Morales, P. M. Rosset, and J. M. Duarte, 'Seed sovereignty and agroecological scaling: Two cases of seed recovery, conservation, and defense in Colombia', *Agroecology and Sustainable Food Systems*, vol. 43, no. 7–8, pp. 827–847, 2019, doi:10.1080/21683565.2019.1578720.

[22] Mesa Nacional Agropecuaria y Popular de Interlocución y Acuerdo, 'Gran paro nacional agrario Colombiano', 2013. Accessed: 14 Jun. 2022. [Online]. Available: https://www.mpdl.org/sites/default/files/131024-paro-agrario-completo.pdf.

[23] Presidencia de la República de Colombia and FARC-EP, *Acuerdo Final Para La Terminación Del Conflicto Y La Construcción De Una Paz Estable Y Duradera*, 2016. [Online]. Available: http://www.altocomisionadoparalapaz.gov.co/procesos-y-conversaciones/Documentos compartidos/24-11-2016NuevoAcuerdoFinal.pdf.

[24] A. P. Cubillo-Guevara and A. L. Hidalgo-Capitán, 'Good living goals. A proposal for the construction of a global transdevelopment', *Alternautas*, vol. 6, no. 2, pp. 71–87, 2019.

[25] B. de S. Santos, *Epistemologies of the South – Justice against epistemicide*. New York: Routledge, 2016.

[26] S. Schavelzon, 'El concepto de *Vivir Bien/Buen Vivir*', in *Plurinacionalidad y Vivir Bien/Buen Vivir*. Buenos Aires: CLACSO, 2015, pp. 181–267.

[27] E. V. Wagner-Medina, A. M. Santacruz Castro, and C. P. Rendón Ocampo, 'Sistema de semilla en Colombia: Consideraciones sobre calidad y agrobiodiversidad', *Estudios Rurales*, vol. 11, no. 22, 2021. [Online]. Available: http://portal.amelica.org/ameli/ ja tsRepo/181/1811955002/index.html.

[28] GOV002, 'private communication', Mar.2021.

[29] Campaña semillas libres, 'Manifiesto por las semillas libres en Colombia', 2012. [Online]. Available: http://www.semillas.org.co/sitio.shtml?apc=f1-1–&x=20158142.

[30] NET007, 'private communication', Oct.2021.

[31] I. Vargas-Chaves, A. Gómez-Rey, and G. A. Rodríguez, 'La invocabilidad de la accion de cumplimiento frente a los compromisos suscritos en el Paro Nacional Agrario de 2013: Un analisis desde la regulacion de las semillas certificadas en Colombia', *Revista Jurídicas*, vol. 15, pp. 28–45, 2018, doi:10.17151/jurid.2018.15.1.3.Recibido.

[32] R. Vera Herrera, 'Convite de saberes de custodios y guardianes de semillas', *Biodiversidad – Sustento y Culturas*, pp. 16–17, 2015.

[33] T. Aguilar Gómez, A. M. García Hoyos, M. García Álvarez, and Campaña Semillas de Identidad, 'Sistema participativo de garantia de la calidad de semillas criollas y nativas para las casas comunitarias de semillas en Colombia – Guía metodológica', 2018. [Online]. Available: www.semillas.org.co/portal/tecnicas/1.4 Sistemas participativos de garantias/cartilla_spg.pdf.

[34] Ministerio de Agricultura y Desarrollo Rural, *Resolución 464*, 2017. Accessed: 14 Jun. 2022. [Online]. Available: https://www.minagricultura.gov.co/Normatividad/Resoluciones/Resolución%20No%20000464%20de%202017.pdf.

[35] GOV001, 'private communication', Mar.2021.

[36] NET009, 'private communication', Jan.2022.

[37] D. Silva Garzon, 'Three logics of action and reaction for the monopolization of seed markets in Colombia', *Revista Colombiana de Antropologia*, vol. 55, no. 2, pp. 9–37, 2019, doi:10.22380/2539472X.795.

[38] C. Hookway, 'Some varieties of epistemic injustice: Reflections on Fricker', *Episteme*, vol. 7, no. 2, pp. 151–164, 2010, doi:10.3366/E1742360010000882.

[39] GOV002, 'private communication,' Sep.2025.

[40] Corte Constitucional de la República de Colombia, *Derechos fundamentales individuales y colectivos de comunidades indígenas – Vulneración por derechos fundamentales a la alimentacion, la seguridad alimentaria y la soberania alimentaria*, 2023. Accessed: 5 Dec. 2025. [Online]. Available: https://www.corteconstitucional.gov.co/relatoria/2023/t-247-23.htm.

[41] Ministerio de Agricultura y Desarrollo Rural, *Anexo Técnico – Lineamientos de Política Pública Agricultura Agroecologica Colombia*, 2024.

[42] AGROSAVIA, 'Plan de acción para la conservación, multiplicación, uso e intercambio de semillas nativas y criollas II Fase 2024–2028', Oct.2024.

[43] Ministerio de Ambiente y Desarrollo Sostenible, 'Minambiente y comunidades del Chocó restaurarán 300 hectáreas en el río Atrato', n.d. Accessed: 8 Feb. 2026. [Online]. Available: www.minambiente.gov.co/minambiente-y-comunidades-del-choco-restauraran-300-hectareas-en-el-rio-atrato/.

[44] AGROSAVIA, 'MinAgricultura lidera rematriación de semillas de maíz nativas para el pueblo Kamëntsá del Putumayo', n.d. Accessed: 8 Feb. 2026. [Online]. Available: www.minagricultura.gov.co/noticias/Paginas/MinAgricultura-lidera-rematriación-de-semillas-de-maíz-nativas-para-el-pueblo-Kamëntsá-del-Putumayo.aspx.

[45] N. Quiceno Toro, *Vivir Sabroso: luchas y movimientos Afroatrateños, en Bojayá, Chocó, Colombia*. Bogotá: Editorial Universidad Rosario, 2016. doi:10.12804/th9789587387506.

5 Venezuelan seed system transformations

This chapter explores the transformation of the Venezuelan seed system over the last 25 years, a process kickstarted by the Bolivarian Revolution, which produced a constitution establishing an eco-socialist model of development that promoted altered state–society relations and more horizontal human–nature relations. These altered relations allowed historically marginalised practices and narratives to be revalued and empowered. In a highly concentrated food system structured as an import-distribution complex to capture petrodollars, marginalised actors found a new voice, connected with transnational food sovereignty social movements, and accessed critical decision-making spaces.

When a seed law started being discussed in 2012, marginalised food system actors connected and mobilised physically and virtually to participate in drafting the text. Seed guardian networks (SGNs) and their allies used Bolivarian Revolution institutions like 'street parliamentarism' to ensure their visibility in the emerging legislation. However, altered state–society relations did not directly translate into altered state–capital relations, as these participative legislative processes kept encountering resistance from confederations representing corporate interests.

In addition to resistance from formal system actors to increased participation of the informal system, pervasive agro-industrial biases in government constantly questioned the validity of AfroIndopeasant agricultural knowledge(s) and practices. Despite these barriers to move the seed system towards a more inclusive model, social movements successfully manoeuvred the legislative process to ensure that historically marginalised actors were represented in the law.

In late 2015, the Venezuelan seed law was approved, becoming a global reference for progressive seed legislation. The law acknowledges the importance of AfroIndopeasant agricultural practices, recognises plurality of knowledge(s) and knowledge-holders, and sets the stage for the emergence of novel institutions that encourage dialogues between technoscience and traditional ecological knowledge.

Regretfully, the Bolivarian model entered a deep crisis after the law was approved, slowing down seed system development under this new set of rules. While social movements continued their work of seed guardianship and exchange, the state faced critical conditions, which stunted the implementation of the programmes outlined in the law.

DOI: 10.4324/9781003715092-8

Implementation of the 2015 law continues today with a backdrop of a multidimensional crisis, capitalist interests pushing to change the law to align with their visions, as well as resistance from traditional leftists pushing for more agricultural industrialisation. The dynamics of this change process can be understood as a series of nested transformative events emerging from previous structuration (See Figure 5.1).

Precedents

Bolivarian revolution

Before going into the particularities of the Venezuelan seed system and its transformation, it is important to dedicate a few lines to the broader stage of the Bolivarian Revolution. This political process was led by Hugo Chávez, a

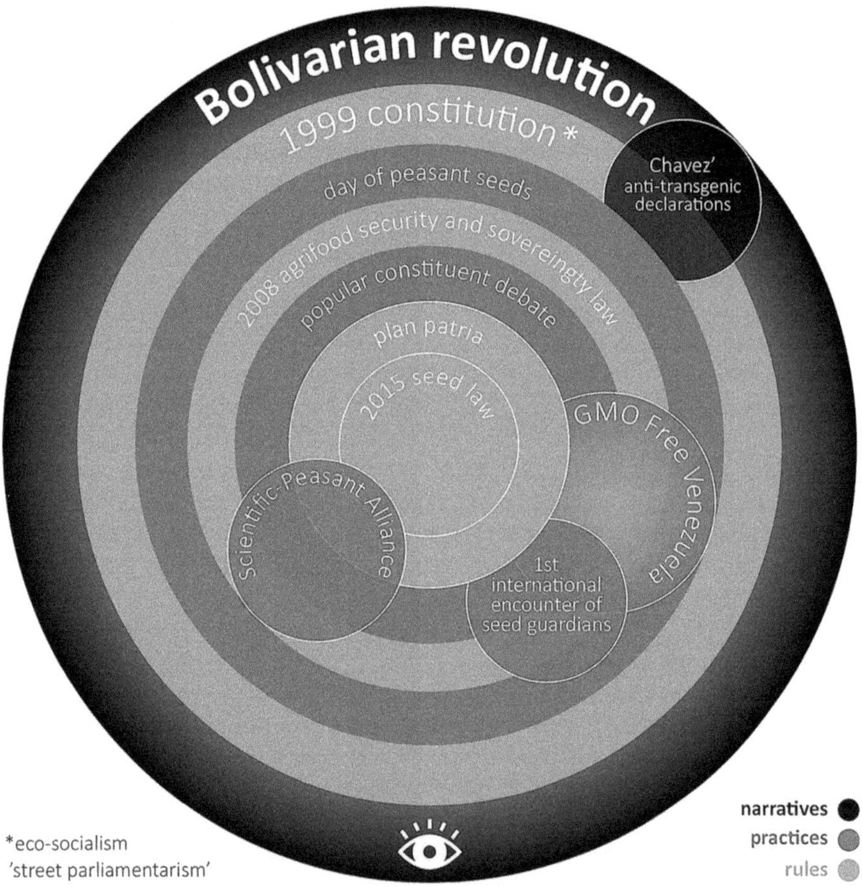

Figure 5.1 A visualisation of Venezuelan seed system transformations as a series of nested fields emerging from previous structuration

military officer who founded a leftist revolutionary movement within the armed forces in the 1970s [1]. The movement was ideologically heterogeneous but aimed for a middle ground between the neoliberal model and radical leftist positions of insurgencies like the communist Red Flag Party.

From 1977 to 1984, Chávez maintained a secret cell in the Venezuelan army, which constantly recruited personnel. After several years of planning, Chávez executed a military coup in 1992 to remove then President Carlos Andrés Pérez from power. Pérez had implemented IMF-supported policies, despite promising the opposite during his campaign. These austerity measures led to widespread protests and looting known as the 'Caracazo', which were met with violent military repression [2].

The 1992 coup d'état failed, but gave Chávez notoriety, since his surrender speech was broadcast to the nation. While Chávez was in prison, another coup was thwarted in late 1992, but Pérez was ultimately impeached in 1993. Chávez was released in 1994 and founded the MVR (*Movimiento Quinta República*, Fifth Republic Movement) political party in 1997 to participate in the presidential election to be held in 1998. Hugo Chávez was finally elected president with 56 per cent of the vote, mainly from the poor and middle classes, reflecting a historical dissatisfaction with traditional parties [3] and the fact that more than half of the population was living in poverty [2].

After his election, Chávez moved quickly to dismantle existing systems of checks and balances, and positioned family and allies from leftist politics and the military in positions of power. This move was widely criticised for establishing a 'Bolivarian bourgeoisie' with little experience in public administration. Despite these shortcomings, the Bolivarian process also opened key spaces for people from social movements in relation to the government, 'producing a more complex arena within institutions allowing new agendas to emerge' [4].

New actors were included in state institutions, while government was encouraged to collaborate more closely with civil society. This kept blurring the boundaries between state and society [3]. The revolution also established novel political and legal tools like 'street parliamentarism', which is 'nothing more than encouraging people through their organisations to propose their own laws, ... it was just as valid [for civil society organisations] as [for] a group of parliamentarians to draft a text for a possible law' [5]. Indeed, one of the first actions of the revolution was to 'build the legal architecture necessary for broad-based popular participation in governance, towards a vision of "protagonistic [sic] and participatory democracy"' [3, p. 654].

In 1999, Chávez called an election to produce a constitutional assembly inclusive of Indigenous groups [3]. The resulting constitution and subsequent legislation share an eco-socialist frame giving more attention to ecological issues and expanding participation through the institutionalisation of 'popular power' [6].

Relevant to food and seeds, the constitution [7] gives special attention to popular cultures constitutive of 'Venezuality' (Venezuelan identity), recognising and respecting interculturality under the principle of equality of cultures (Article

100). It guarantees and protects the collective intellectual property of Indigenous knowledge(s), technologies, and innovations, prohibiting the patenting of their genetic resources and associated knowledge (Article 124). It positions the state as promoter of sustainable agriculture as the strategic basis for overall rural development, guaranteeing food security and self-sufficiency by prioritising and developing internal agricultural production (Article 305). Finally, it enshrines the rights of Indigenous peoples (including to their intellectual property) and attempts to progressively eliminate large estates (Chapter VIII, Article 307).

The new constitution was approved in December 1999, increasing protections for Indigenous peoples and the environment, expanding state benefits, but also increasing presidential powers and reshuffling existing institutions, giving more power to the military. Chávez was re-elected in the year 2000 as part of elections intended to legitimise a post-constitutional government. During his second presidential term, tensions with the United States of America (USA) began to rise due to Chávez's condemnation of the invasion of Afghanistan. Additionally, the president began to meddle with the autonomy of the state-run oil company PDVSA (Petroleos de Venezuela) by replacing managers and executives with political allies. Chávez then used PDVSA to take control of the national oil industry by establishing majority ownership in privately owned companies.

These measures led to a better-organised opposition that resisted a perceived 'Cubanisation' of Venezuela and increased authoritarianism from Chávez and his allies. This ultimately resulted in a coup attempt against the Bolivarian government in 2002 and a recall referendum in 2004 to remove Chávez from power (both unsuccessful). Chávez went on to serve three presidential terms until his premature death due to cancer in 2013, as his fourth presidential term was starting.

His then vice-president Nicolas Maduro was elected president in April 2014, facing renewed disapproval of the Bolivarian Revolution and an intensification of mass protests against it. For some, this signalled the decline of this political model [8]; for others, these were a symptom of the revolution's transformative potential, as resistance appeared almost exclusively in middle- and upper-class areas, led by a largely young, urban, and white crowd [2]. The Bolivarian model continues to this day. However, in January of 2026 the government of the USA captured President Nicolas Maduro after a brief military incursion into Venezuelan territory. This opens a new chapter of instability for the country, as there is increased uncertainty on the continuation of the Bolivarian Revolution.

Contours of the Venezuelan food system

History of the import-distribution complex

The 1930s are critical in the history of Venezuelan food systems. The confluence of the 1929 USA stock market crash and the development of the oil industry in Venezuela resulted in most wealthy Venezuelan families shifting their capital from agro-export businesses to oil exploitation concessions. This was paired with an acceleration of proletarianisation and urbanisation, removing capital from rural areas and increasing urban poverty [2].

The 1930s also kickstarted the process of agricultural modernisation, as the increase in oil revenues translated into imports of supplies, machinery, and agricultural implements [6], as well as the establishment of strong relationships between agribusinesses and the state [3]. During this decade,

> owners of the former agro-export complex were able to take advantage of its existing infrastructure, the inflow of oil dollars, and the newly acquired purchasing power of Venezuela's emerging middle class to shift from exports to imports, giving rise over time to a powerful agri-food import-distribution complex inextricably linked to petroleum extraction [2, p. 5].

Subsequently, the arrival of the green revolution in the 1950s consolidated economic competitiveness as the guiding principle for agriculture, leading to planned efforts by technical and political elites to substitute traditional food production with industrial agriculture [6]. These efforts made the Venezuelan agricultural sector the most mechanised in Latin America at the time, combining the use of specialised machinery, improved seeds, and agrichemical inputs [2].

The 1950s also brought the first seed production programmes and certification standards, implemented through public–private partnerships, mainly benefiting large-scale companies [3], and aimed at expanding the frontiers of mechanised agriculture [6]. In 1961, an institution dedicated to seed certification was established (SECERSEM) and the first certification regulation (INV-71) was enacted by the Ministry of Agriculture and Livestock. By 1986, general seed standards (NORGESEM) and a National Seed Service (SENASEM) were created [6]. However, despite all efforts to strengthen the Venezuelan food system, the sector was still relatively weak, since it operated in an import-focused rentier society (especially in the neoliberal 1990s) [9].

Venezuela's agro-industrial sector is still today 'oriented towards importing food to capture foreign currencies that are centralised by the oil industry' [4]. This was exacerbated after 2003, as oil prices skyrocketed, leading to more caloric dependence on outside sources [10]. Venezuela's import-based system has an 'oligopolistic structure and is articulated with the transnational agri-food industry … [with Venezuela] as the recipient of [global] agricultural surpluses' [11, p. 4] including supplies and technologies.

Historically, Venezuelan agro-industrialists have not aimed to develop an agro-industrial sector, but to 'capture the state's foreign currency for their import-export businesses and to expand their capital outside of Venezuela' [5]. This has resulted in the national agricultural industry having minimal participation in the internal economy, where just a few big food-producing companies control much of the diet of urban Venezuelans [5]. However, this does not mean that transnational corporate interests are absent, as they are strongly represented by national associations or confederations [3].

Seed system generalities

This macro framing of Venezuelan agriculture trickles down to seeds as one of its primary inputs. As Buitrago emphasises, 'to understand seed production, it is necessary to understand the developmentalist logic of extractivism in Venezuela, especially oil, since this logic has been the organiser of ways of life and social relations in the country at all scales' [9, p. 78]. Aligned with this logic, the formal seed system in Venezuela has been and continues to be highly dependent on imported seeds produced by private companies and international germplasm banks [3].

At the beginning of the 21st century, there still wasn't enough national seed production to fulfil market needs, and locally produced hybrid seeds responded to outdated quality standards [11]. In addition, quality standards for imported seeds were and still are less restrictive than those for nationally produced seeds [6]. The 2005 National Seed Plan attempted to mitigate these issues but achieved mixed results. This plan was the 'central seed research and production policy … aimed at developing novel varieties that could be used by the majority of peasants' [4].

More broadly, the process of capitalist industrialisation of agriculture in Venezuela encouraged the production of a very small number of agro-industrial products, like cereals, textiles, oilseeds, and sugar cane, while ignoring products historically associated with self-subsistence, like grains, legumes, and tubers [12]. Adding to this, only 11 of the 79 vegetable products in the Venezuelan food system use certified or supervised seeds despite all existing formal structures [13].

This evidences the importance of non-certified farmer seeds in the Venezuelan food system, which have historically been rendered invisible in national laws and regulations [6], an invisibilisation that has led some authors to state that the seed system is ordered in a racialised manner, with AfroIndopeasant seeds being 'systematically displaced by agribusiness, revealing the supremacist logic of [the] corporate model' [11, p. 21]. This invisibilisation is just another example of the pervasive agro-industrial biases in state institutions and beyond.

For confederations representing corporate interests in Venezuela, 'the issue of seeds is concentrated in very few [varieties] … that are part of industry: soy, corn, and everything that is part of the processes of animal production' [5]. Confederations work closely with the National Institute of Agricultural Research (INIA, *Instituto Nacional de Investigaciones Agrícolas*) to develop new varieties or register breeder rights in what represents an 'important state subsidy for the private seed sector' [3, p. 656], additional evidence of a state that responds almost exclusively to the interests of agribusiness.

Relevant legislation

Post-constitutional legislation transformed the food system in Venezuela, setting the stage for the 2015 seed law. It is important to mention that Venezuela is not part of the International Union for the Protection of New Varieties of Plants

(UPOV), so it is outside of transnational systems for plant variety protection, allowing more flexibility in national laws.

Very soon after the approval of the Bolivarian constitution, the Law of Land and Agrarian Development (2002) [14] set a legal framework for agrarian reform and recognised the *conuco*[1] as a historical source of agro-biodiversity (Article 19). This law privileged collective organisations in agricultural production (Article 4, 5), protected ancestral agricultural techniques, and promoted research about them (Article 17, 19), also establishing measures against large unproductive estates.

In 2008, several laws were approved moving the Venezuelan food system towards food sovereignty and agroecological framings. These are: the Law of Integrated Agricultural Health [15], the Law of Management of Biological Diversity [16], and the Organic Law of Agri-food Security and Sovereignty [17].

The Law of Integrated Agricultural Health includes provisions for the development and promotion of agroecology and expanding popular participation in agricultural health. Crucially, it positions agroecology as a pathway for the transformation of the national socioeconomic model. The Law of Management of Biological Diversity aims to preserve and re-establish ecological balances that would allow humans and other living beings to live in harmony with their environment. It also established the need to prevent negative impacts of GMOs, and to include traditional knowledge related to biodiversity (Article 2).

The Organic Law of Agri-food Security and Sovereignty establishes agri-food sovereignty as the right of a nation to define and develop agri-food policy according to its own needs. This includes a recovery of traditional agricultural practices and technologies with a focus on the sustainable development of future generations (Article 12), as well as the development of new technologies (Article 15). It also promotes the fair distribution, exchange, and commercialisation of food products, including barter and other forms of non-monetary exchange (Title III). Finally, it establishes participatory planning instances in multi-scale agrarian assemblies inclusive of peasant producers (Chapter 2, Second Section).

It is important to mention here the 'All Hands to Sowing' programme implemented since 2009 by the Ministry of Popular Power for Education. This initiative incorporates agroecology curricula into the Venezuelan education system as a transition strategy from an agrichemical to an agroecological model. Curricula include: agroecological soil management; water and agroecology; ecological control of insects, disease, and spontaneous vegetation; and seeds. This last item aimed to define seeds as of strategic importance for food sovereignty and includes thematic contents such as seed selection, importance of peasant seeds, basic seed quality tests, and socialist seed distribution, among others [18].

Additional contours

There are four additional variables of Venezuelan food systems worth mentioning: (i) AfroIndopeasant agriculture plays an important role in national food production; (ii) Venezuela has avoided the introduction of transgenics;

(iii) there are many contradictions in the implementation of legislation and the Bolivarian model; and (iv) food sovereignty social movements have grown considerably during the Bolivarian Revolution.

First, traditional Indigenous, Afro-Venezuelan, and peasant agriculture plays an important role in feeding the country [19]. The *conuco*, an ancestral agricultural institution kept alive by small-scale farmers today, is responsible for a large amount of national food production. It is important to situate this phenomenon historically, as food exports during the colonial period were produced by colonial elites in large plantations using mainly slave labour, while labourers used communal plots for their subsistence [2]. This persists to this day, with the organisation of agro-industry on mechanised large estates, while AfroIndopeasant agriculture occurs on smallholdings still generally referred to as *conucos.*

Second, despite Venezuela having a high concentration of land tenure with a predominance of large estates, 'it was always much more profitable for capital to appropriate oil income via imports, and not via the occupation of the land with technologies that are … expensive and that imply [capitalists] have to work' [4]. This means that Venezuela is 'outside of the frontiers of transgenic monoculture expansion, because of its role as oil provider' [4].

In addition to this structural issue, then President Chávez issued what could be interpreted as an informal prohibition of transgenics in 2004 [6]. During his recurring televised programme *Aló Presidente*, Chávez stated that 'here in Venezuela we are working, and we have already banned a trial that they wanted to introduce here with transgenics, and we are putting up the respective barriers at national level to transgenics, which do a lot of damage to agriculture and above all to the sovereignty of our peoples' [20]. After 2004, the Institute of Integrated Agricultural Health ensured that seeds entering the country were not transgenic [4], but presently no law explicitly forbids the use of transgenics in the country.

Third, there are many contradictions in Venezuela's food system that have only increased in recent years due to the systemic crisis of the Bolivarian model. Worth mentioning are:

- i Conflicts between national and supranational legislation, as exemplified by the unconstitutional recognition of plant breeder rights as part of Venezuela's entry to the Andean Community of Nations (CAN) [6].
- ii A decrease in per capita agri-food production, while dependence on food imports increased throughout the Bolivarian Revolution (data until 2014) [21]. These imports are often part of state-run food programs allocating state oil revenues to the private agri-food import-distribution complex, effectively subsidising powerful companies [2].
- iii Use of humanitarian aid due to the current economic crisis has led to 'new dynamics of reinforcing control over … seed systems' [11, p. 21].

As Buitrago states, despite the Bolivarian Revolution, 'the lack of a diversified productive matrix makes Venezuela vulnerable to an onslaught from the hegemonic world order of corporate agribusiness' [9, p. 79]. As other Venezuelan researchers agree, these issues 'are symptomatic of the fundamental structural problems of a food system based on imports and dominated by the interests of capital, through deep [capital–state] alliances that have been forged over the course of history' [2, p. 22].

Finally, the last 20 years have seen an increase in the number and strength of social movements concerned with food sovereignty and agroecology [3]. This has to do with the framing of Venezuela as an eco-socialist space where an ongoing revolution moves towards alternatives, but also with the global spread of these concerns, as exemplified by organisations like 'La Via Campesina', which coordinates organisations globally around issues of peasant and family agriculture and food sovereignty.

Over the last two decades, important spaces for seed production, knowledge production, and training emerged, playing an important role in configuring a seed movement in Venezuela [4]. This seed movement was propped up by collectives like PROINPA[2] in Merida (focused on the recovery of native potato varieties), *Sin maíz no hay país* in Guanape (focused on corn), and Monte Carmelo in Trujillo.

This movement has been promoting transformations in the Venezuelan food system beyond formal institutions, by strategically using leverage points in their interactions with the state as political opportunities emerge [3]. The following section reviews the process by which the 2015 seed law was approved from the perspective of these social movements. This law was highly progressive as it is arguably the biggest historical victory for marginalised food producers in Venezuela.

Path to the 2015 Seed Law

National Peasant Seed Day

As a product of articulations between native seed guardians and the National Institute of Agricultural Research (INIA), a day to celebrate peasant seeds was instituted in Venezuela. Since 2005, diverse seed system actors gather in the village of Monte Carmelo to share knowledge, exchange native seeds, and resist transgenics and capitalist agriculture. Monte Carmelo has become the 'meeting place for all small, large, new and traditional seed initiatives' [4], [5]. Moreover, this space and the experiences it has enabled have become a 'benchmark for the issue of peasant seeds in Venezuela' [4], being a 'very important part of the process of generating the [2015] seed law proposal' [5].

Two instances of the National Peasant Seed Day (*Día Nacional de la Semilla Campesina*) will be explored in this section, its first and seventh iterations. The first iteration produced the declaration of peasant seeds, which evidences the goal of solidarity between farmers and seeds, cementing the non-Modern foundations of this space, which would inform the transformation process to come. By its seventh year, this space had consolidated as a powerful political force in opposition to capitalist agriculture. It readied for mobilisation, using the participatory legal frameworks of the Bolivarian Revolution to draft a proposal for the 2015 seed law, ultimately transforming the rules of the national food system.

A non-Modern seed is planted

In 2005, as part of the first National Peasant Seed Day, participants signed a declaration of peasant seeds [22]. The text was written by peasant farmer Gaudy María García and exemplifies the intention of these networks to acknowledge non-human perspectives. The declaration is written as if seeds were the authors, using a first-person perspective to communicate agency and personhood. Seeds are presented as being part of complex systems in relation to insects, birds, and micro-organisms. There is an acknowledgement of their emergence from 'mother earth'. There are also many mentions of solidarity between seeds and peasant farmers, as well as with Venezuelan food consumers. The text mentions mistreatments of seeds over the last century, referring to the use of agri-chemicals, and calls for education regarding their importance, the creation of community-managed seed banks, and AfroIndopeasant-led care.

Moment zero

Seven years after the signing of the declaration of peasant seeds, the first international meeting of seed guardians was held in Monte Carmelo. More than 300 people representing 100 organisations attended the event, including people from Argentina, Bolivia, Brazil, Colombia, Ecuador, El Salvador, México, Paraguay, Uruguay, and Venezuela [23]. The meeting was organised under the slogan: 'Guard the knowledge of the best seeds of the best blossoms to guarantee anti-capitalist food independence'.

The agenda included an exchange of rituals and *pagamentos*[3] associated with seeds and the *conuco*, as well as seed exchanges. In fact, one of the requirements to attend was to bring local seeds to barter with other participants. One of the objectives of the meeting was to promote dialogue between guardians in three roundtables about: (i) strategies and spaces for seed production, custody, and multiplication for the preservation of seed biodiversity and the *conuco* economy; (ii) traditional protocols, legislation, and public policies for the preservation of seed biodiversity; and (iii) pedagogical and communication strategies for the preservation of seed biodiversity.

The meeting produced a declaration framing seeds as the patrimony of humanity and not the property of a few corporations while rejecting the use of transgenics and agrichemicals. The declaration framed local knowledge-holders (*maestros-pueblo*) as leaders in the transformation of the economic system. It set out to overcome understandings of knowledge production as exclusive to universities and official institutions, attempting to 'lift the veil' of capital that hides the sacred bond between food, medicine, and seeds. The declaration was oriented around four axes: (i) production, custody, multiplication; (ii) legislation and public policies; (iii) research and education; and (iv) communication [24].

Axis (i) demanded preservation of Indigenous and peasant seeds, as well as water and *conuco* culture. It declared a continuous resistance to 'Capitalism, Imperialism, Large estates, Monoculture, Transgenics, Agrochemicals, and the Gospels according to empire and Wars' [25, p. 3]. It established a commitment to self-governance to maintain seeds in the hands of popular power, by creating seed houses and fostering seed exchanges.

Axis (ii) reinterpreted the ecological framing of the 2013–2019 government development plan to propose a new national objective: 'Recognise and respect the rights of mother earth as the basis of *Buen Vivir* to overcome capitalism and the overarching civilisatory model' [25, p. 3]. For participants, this would translate into the other two objectives, the protection and promotion of Indopeasant seeds and associated knowledge, as well as maintaining Venezuela as a country free of transgenics.

Crucially, the 2012 National Peasant Seed Day celebrations coincided with news that a seed law was being discussed in the National Assembly. Participants 'didn't know much about the contents of the law, so [they] agreed that people living closer to Caracas, which is the centre for decision-making, should monitor what was happening and try to influence the elaboration of that law' [4]. In line with this, axis (ii) included a demand to participate in the construction of the new Venezuelan seed law [24].

Axis (iii) sought to 'rescue' Indigenous and peasant knowledge, as well as to recover traditional food and medicinal memory. It advocated for the inclusion of this knowledge in schooling at all levels and committed to developing participatory and emancipatory seed research. Finally, axis (iv) established the Venezuelan National Seed Guardian Network, linking it to other similar initiatives in Venezuela and Latin America. The network also committed to develop a large communications campaign to publicise the 'emancipatory capacities of Indigenous and peasant seeds' [25, p. 5].

The 2012 National Gathering on Peasant Seeds also led to the launch of the Campaign for a GMO-Free Venezuela (*Campaña Venezuela Libre de Transgénicos*). The campaign grouped a diverse range of actors under a common banner beyond differences in existing agendas to intervene in the seed law process [3]. It enabled an encounter between two spheres, urban movements (young people, intellectuals, environmental activists, diversity activists, etc.) and rural/peasant movements. These stakeholders shared environmental, agri-food, and ecological concerns converging in the issue of transgenics (concerns that were exacerbated by social media at the time).

News of the proposed seed law led these seemingly disparate actors to 'move, [or] at least to ask: "What's going on? Where does it come from? Who's proposing this?" … to ask what was happening and make a counterproposal' [4].

Adding to these emerging coalitions, the Venezuelan Congress on Biodiversity was also held in 2012, with a wide range of actors focusing on rescuing traditional seeds. Interestingly, around this time the Colombian documentary *9.70* (discussed in Chapter 4) was being broadcast on national television. This led informal(ised) seed system actors to 'raise the alarm, [as the documentary] made it easier for us to understand the risks … deriving from the implementation of laws that are basically anti-peasant' [4]. Understandably, the dynamism of the seed sector throughout the year has led many to frame 2012 as 'moment zero' for the 'battles around the seed law' in Venezuela [3, p. 659].

2013–2015: Seed Law battles

Death of Chávez, a call to attention

Hugo Chávez, the figurehead of the Bolivarian Revolution, died in early 2013. His passing opened a space of uncertainty, despite leaving behind the 1999 constitution. For food sovereignty social movements, his death was a wake-up call, infusing movements with a 'new sense of energy and urgency, especially with the risk of clashing institutional orientations of the state shifting further away from radical agrarian politics and more towards depoliticised technological solutions' [3, p. 660].

The movements also reflected on the role Chávez had played in activating popular power and creating the structures to mobilise it [3], an arena that network participants were keen to take advantage of, especially since by this point they felt empowered after a decade of multiple organisational efforts consolidating in the country [5]. Accordingly, the Campaign for a GMO-Free Venezuela redoubled its efforts.

By then, the campaign had grown considerably, linking both rural and urban social movements, united mainly by a vision of popular control of the food system, beginning with seed sovereignty [5]. The campaign's vision also included a resistance to industrial agriculture, identification with transnational food sovereignty principles, and an affinity with place-based agroecology [3]. The campaign made its first public appearance in May 2013 as part of the global March against Monsanto, an event they leveraged to raise awareness of the discussion being held around a new seed law. On that day, the campaign collected signatures on a statement that was sent to the National Assembly, which opened a space for participation in the subcommittee discussing the upcoming law [3].

Drafting a new seed law

The Campaign for a GMO-Free Venezuela started attending meetings of the Permanent Subcommittee on Agri-food Development in the National Assembly, joining representatives of several ministries, the Ombudsman's Office, and the Bolivarian University (Universidad Bolivariana). The campaign wanted to audit and to demand popular participation in the creation of the law, commenting on a first draft, which had been developed by the technical commission of the National Assembly [24].

From these meetings, the subcommittee organised popular constituent debates to discuss the law. Eventually, the debates produced an amended draft, which was leaked to the campaign by an ally inside the National Assembly. This draft showed that input from social movements was absent, reflecting existing agro-industrial biases. It demonstrated that 'many government officials believed that peasant-based agriculture was politically correct but not technically correct' and that, despite participation by the movements, 'altered state-society relations do not automatically translate into altered state-capital relations' [3, pp. 657–660].

As a response, GMO-Free Venezuela organised protests, social media campaigns, and signature collections, and strengthened its alliances with sympathetic government actors. This second round of activism led to an agreement with the National Assembly that the law would be developed with social movement participation through a mixed process of legislative and popular power. That is, people would create the law beyond a mere process of consultation to validate existing proposals [3].

Recalling the political and legal tools created by the government under Hugo Chávez, this process was framed by the movements as an exercise in 'street parliamentarism'. Popular organisations leveraged the opportunities for participation that such institutions provided in an attempt to effectively take over the seed law process [5]. In late 2013, the campaign met again in Monte Carmelo for National Peasant Seed Day, where the first meeting to discuss the law occurred. This kickstarted two parallel strategies, a series of public consultations by the National Assembly, and a popular constituent debate organised by social movements [3].

The process spilled over to the public, moving beyond interested parties to become a national discussion. As the movements increased their participation in the drafting of the law, the seed industry took to the press to reassert their positions. The Campaign for a GMO-Free Venezuela saw this initially as a threat, but quickly realised that it favoured them, since the topic was on everybody's lips.

The wider public knew that discussions were being held, that issues with transgenics were part of the debates, and that a seed law was in the works. This awakened broad solidarity, turning the street parliamentarism process into a 'movement with national and international presence, [with] statements of solidarity from other movements, … signatures from academics, and … more initiatives supporting it' [5]. The social movements also deployed other narrative strategies to sway public opinion, namely, referring to the president of the subcommittee – and proponent of the law – as 'Deputy Monsanto', a moniker that stuck throughout the whole drafting process [5].

This process shifted GMO-Free Venezuela from a defensive to an offensive position. The drafting process was framed as a confrontation with a clearly delineated enemy: corporate interests protected by the state. The movements would develop a draft of the seed law to counterbalance the initiative presented by the National Assembly. This coalesced all social movements involved. As expressed by a participant, it 'unified the people a lot ... it's not that we're going to negotiate a law! We're going to work on a law that opposes this other option' [5].

In this space, discussions moved from what should be resisted to proposals of what the law should be for, mainly making visible and revaluing AfroIndopeasant seed knowledge(s) and practices, as well as the *conuco* as a form of production [3]. One of the activists involved refers to this process as having two stages:

> First, the movement was weak and appeared like spoiled youngsters who didn't want transgenics. [Technicians mobilised] their pseudo-technical discourses [to frame the movements as] against global technical progress, cutting-edge technologies, and whatnot. But later, as street parliamentarism and agency were really taken up by the people of the street, especially from rural streets, from the fields ... they changed things, they organised their own consultation systems from town to town [5].

It was through this shift that the movements decided to move beyond commenting on existing drafts and produced their own version of the law 'article by article' [3, p. 663]. This popular debate ended up taking three years, as workshops were held across the country to gather people's opinions. These spaces allowed for a deeper 'articulation of different experiences working on the issue of peasant seeds at the national level, allowing [the campaign] to have a broad picture of the diversity of experiences, struggles, perspectives, types of seed, types of technologies, and actors' [4].

Plan de la Patria 2013–2019

A crucial element that energised the Campaign for a GMO-Free Venezuela and opened more spaces for it in seed law discussions was the approval of the Plan de la Patria 2013–2019 (Plan of the Fatherland 2013–2019). In 2013, President Nicolas Maduro presented this six-year economic and social development plan for Venezuela, which had been developed by Hugo Chávez before his death.

The plan was approved by the National Assembly in early December, with the vision of

> turning Venezuela into a national power in the social, economic, and political spheres, highlighting the idea of national power associated not only with the economic and productive component, but also with the social dimension, as well as a power supported by transformative education. *All in service of a vision of a relationship between humans and nature in which the latter is considered as a subject and not as a mere commodifiable object* [emphasis added] [26, p. 5].

Under this vision, the plan [24] focused on some of the following objectives:

i Continue building Bolivarian socialism of the 21st century in Venezuela, as an alternative to the destructive and savage system of capitalism.
ii Contribute to the development of a new international geopolitics in which a multicentric and multipolar world takes shape, allowing balance for the universe and guaranteeing planetary peace.
iii Preserve life on the planet and save the human species.

As part of these objectives, the plan included (among other points):

- A strengthening of popular power to be mobilised towards emancipation; achieving food sovereignty by eradicating large estates and increasing peasant access to agricultural inputs and popular power; and promoting 'diversified production models, based on family, peasant, urban, peri-urban, and Indigenous agriculture, recovering, validating, and disseminating traditional and sustainable production models' [26, p. 10].
- Consolidating a transformative, diverse, and creative scientific, technological, and innovative style geared towards solving the needs of Venezuelans under a socialist ethics; promoting the inclusion and *Vivir Bien* (good living or living well) of Indigenous peoples, consolidating a vision of heterogeneity and ethnic diversity, with respect for and participatory inclusion of minorities and native peoples; and 'dismantling and fighting against international schemes that promote the commodification of nature, environmental services, and ecosystems' [26, p. 25].
- Designing a national adaptation plan that would allow the country to prepare for climate impacts as a 'consequence of the irresponsibility of industrialised nations, polluters of the world' [26, p. 26]; and, finally, counteracting the production and valorisation of cultural elements and historical narratives generated from the dominant neocolonial perspective.

As is evident from the text, the Plan de la Patria 2013–2019 doubled down on the eco-socialist framing of the Bolivarian Revolution. It situated the country in a cataclysmic climate crisis produced by rich nations and narratives of nature as a commodity. Of special relevance to seed systems, it positioned food sovereignty and decolonial knowledge(s) as key elements for achieving post-capitalist wellness. Despite the plan's inclusion of oil exploitation as the engine for development, it delineates a clear direction towards post-Development. This macro-level framing of Venezuela and its institutions allowed seed movements to validate their narratives and practices to continue the push to transform national seed system rules.

2015: Last push to transform seed systems

Over the period of 2013–2014, seed system stakeholders in Venezuela had engaged in public consultations and a popular constituent debate to draft a new seed law. However, it was evident that these processes were being held in parallel, with little cross-pollination between stakeholders. In early 2015, four different versions of the law were presented in public consultations: one from the National Assembly, one from INIA, one from FEDEAGRO (the main confederation of mid- and large-scale producers in Venezuela), and one from the Campaign for a GMO-Free Venezuela [3].

Several points of contention emerged when comparing these drafts. Some stakeholders pushed for more state control and financing, others wanted more intellectual property protections and promotion of genetic engineering, while the campaign kept pushing for the inclusion of AfroIndopeasant seed practices. Social movements problematised the contradictions between intellectual property rights (IPR) and the rights of farmers, particularly the irrelevance of IPR in traditional collective forms of production by small-scale farmers based on care systems [6].

In line with this perception of incompatibility, the campaign's initial proposal argued for two different systems (formal and informal), each with its own law. This would allow for recognition of new diverse subjectivities in the system (i.e., AfroIndopeasant perspectives), while also acknowledging those already recognised and validated by certification and quality control schemes [3].

As the last trimester of 2015 approached, social movements organised the VI Venezuelan Congress of Biological Diversity. This space was the product of new government and social movement partnerships favoured by allies of the Campaign for a GMO-Free Venezuela in the National Office of Biodiversity (ONDB). The theme of the event was 'seeds for life and for food sovereignty', and the final declaration, signed on October, urged seed system stakeholders to enact the seed law as proposed by popular power [27].

The recently formed Ministry of Eco-socialism and Water (with oversight over the ONDB), INIA, the National Assembly, and the Campaign came together after the event to develop a final proposal for the seed law based on consensus [3].

The campaign shifted its position on having two separate laws, agreeing that formal and informal seed systems should operate under the same law. This change stemmed from recognition that producers used both systems simultaneously, but, more importantly, that the uniqueness of the law comes in part from reflecting

> both the reality of the here and now, as well as the aspirations for a radical transformation of the food system based upon principles of food sovereignty. That both short-term and longer-term visions are reflected in the law, as well as mechanisms for the attempted transitioning from reality to vision [3, p. 665].

By late October of 2015, the race was on to finalise a coordinated draft of the seed law and get it approved before year's end. There was a sense of urgency from the GMO-Free Venezuela Campaign, since in 2016 an opposition majority would be in control of the National Assembly. During one of the last parliamentary sessions of the year, the law was approved, thanks in part to the Chavista majority in power [4]. As some researchers have recounted, the law almost wasn't approved; it required a 'groundswell of mobilisation from below, together with critical openings within the state' [3, p. 649]. The victories of social movements have also been attributed in part to having an 'irreverence' for the law, an attitude that along with Bolivarian participatory institutions enabled a perception of the state as 'hackable' [3].

The approval of the law was celebrated by social movements as a great achievement, as was recounted by one of the proponents:

> In that semicircle, on those balconies, people from the movements … with the strength [of their] work. The formation of … new self-esteem, gained by [their] seeds and [their] knowledge, defending and asking for the approval of the seed law … well, [it was a] great party [5].

Indeed, the victory achieved was no small thing, as evidenced in the final text of the law, examined below.

2015 Venezuelan seed law

This section will focus on those aspects of the law that promote the inclusion of historically marginalised actors, who, despite playing a key role in national food production, were historically absent from its legislation. As seed guardian Pablo put it:

> The importance of this new law is that it will allow us … the exchange of experiences and participation, which we will all have equally. Here in this law, we are not going to be excluded, we are going to have full participation in all the decisions that have to do with the production of food for our country, for our people [28, p. 99].

It is important to emphasise that inclusion of informal(ised) actors does not mean exclusion of formal system actors. The law acknowledges the existence of two seed systems: (i) Industrial seeds subject to certification; and (ii) AfroIndopeasant and local seeds. The law sets out to strengthen the national seed system from an agroecological socialist vision, prioritising national seed production, and valuing AfroIndopeasant and local seeds, while resisting seed patents and prohibiting transgenics from entering the country [29].

The law aims to encourage a transition from conventional food production (monocultures, agrichemicals, industrial seeds) towards sustainable agroecology. It seeks to protect agrobiodiversity through the production of AfroIndopeasant and local seeds. Crucially, it revalues and re-legitimises local, traditional, and ancestral knowledge(s), wisdom(s), beliefs, and practices. To implement all the above, it directs the 'organisation and planning of public policies in function of different scales of production, differentiating policies destined for family agriculture or pluricultures in micro-spaces of production, from those of big producers' [29, para. 9].

The law recognises seeds as living beings that are constitutive of mother earth and, as such, are subject to rights and rules on the preservation of life on the planet and the conservation of biological diversity. Seeds are also framed as a common good of the public interest and the public domain. More broadly, the law is framed under principles and values that include: (i) fighting against the poverty and exclusion produced by capitalism and neocolonialism; (ii) practising equity, inclusion, emancipation, gender equity, social justice, *Buen Vivir*, among other values; (iii) recognising the interculturality present in the seed; (iv) promoting solidary seed exchanges and free access to seeds against intellectual property and patents, thereby defending AfroIndopeasant and local seeds from biopiracy; and, finally, (v) promoting agricultural knowledge exchanges [29].

The law recognised popular power by establishing bodies like the 'Popular Council for the Safeguarding and Protection of Local, Peasant, Indigenous and Afro-descendant Seed'. It created free seed licences to protect local seed knowledge from biopiracy and appropriation via patents. It denominated *maestres pueblo*,[4] the people who create, possess, bear, and transmit oral, popular, and community tradition, as well as AfroIndopeasant knowledge(s), practices, and beliefs on biodiversity [29].

The role of *maestres pueblo* is part of the broader recognition and revitalisation of traditional AfroIndopeasant and local knowledge in the law [6]. This includes a commitment to the *Buen Vivir* of AfroIndopeasant and local populations, and a promotion of the *conuco* as a sustainable mode of production. These recognitions are paired with the promotion of production, distribution, circulation, exchange, and consumption of AfroIndopeasant seeds, the validation of participatory quality assurance systems, and seed collection and storage centres (all under principles of co-responsibility).

It is important to reiterate that these provisions for AfroIndopeasant seeds do not exclude industrial seeds that are subject to certification. The law also creates a National Seed Commission to oversee the national seed registry (RENASEM), encourages the creation of germplasm banks, and promotes seed technological innovation [29]. The law distinguishes research on certifiable seeds and traditional knowledge(s) of seed improvement, but, importantly, it creates a third space of co-responsible participatory research, which would evolve into an institution to be named the Scientific-Peasant Alliance.

The passage of the 2015 seed law was one of the final acts of the Bolivarian-majority National Assembly [3]. The law granted recognition to long-standing agricultural practices that have been an essential part of the Venezuelan food system. As expressed by one seed activist,

> what we did was collect those practices and have them recognised by law so that they were not illegal or outlawed, which is what other laws

> do, they begin to criminalise traditional practices. We made them part of the seed system and on par with everything that has to do with certified seeds [4].

For seed guardians, this is an extraordinary victory, especially in a space with so many interests and a historical trajectory that has favoured the elites. For the social movements, 'all these policy programs in favour of agroecology, food sovereignty, and peasant seeds were the result of a process [of resistance] that is 500 years in the making' [4].

The spirit of the law is to acknowledge Modern biases concerning knowledge, technology, and human–nature relations, to open up Transmodern spaces that aim to close the divide between the Modern and non-Modern. However, the implementation of such an ambitious project faces countless obstacles that produce more questions than answers regarding the integration of those worldviews in food systems. The next section describes the multiple crises faced by the Bolivarian Revolution after the approval of the seed law, and the continued efforts of SGNs to maintain recognition and strengthen their practices.

Bolivarian Revolution in crisis (2015–present)

While some people might argue that Venezuela has been in crisis since the rise to power of Hugo Chávez, it is clear that the drop in oil prices of 2015 was the key moment that plunged the country into an emergency. Since 2015, Venezuela has been suffering from hyperinflation, an increase in morbidity, hunger, and crime, and a scarcity of medicine and other essentials. Other issues include fuel shortages, institutional deterioration, and currency devaluation [11]. The situation is so complex that it resulted in the biggest mass migration in Latin America's modern history. By some estimates, around eight million people have fled the country to find better opportunities, mainly in other Latin American countries [30].

Impacts on the food system

In a heavily import-based food system, the crisis impacted seeds in two particular ways: (i) there was a considerable reduction of imported seeds in the Venezuelan market; and (ii) when these seeds emerged, they were out of reach for the average farmer since they were priced in US dollars. This led to widespread recovery of native seeds and people organising networks of seed growers. Farmers occupied land no longer used for industrial agriculture to expand the *conuco* model, growing vegetables, old varieties of peas, and wheat, among other crops. However, the crisis also left some people trapped in a logic of scarcity, increasing aid dependency, and preventing farmers from fully developing their productive capacities [4].

In recent years there has been a reconfiguration of the forces of agribusiness, with chemical inputs again becoming available, and hybrid and probably transgenic seeds circulating in Venezuela. Between 2018 and 2020, new informal contraband markets for certified seeds emerged, selling imported seeds on Instagram or WhatsApp [11]. Some of these seeds do not align with existing national certification standards, so there is no way for consumers to complain if they are of poor quality. One of the commitments of the state was to monitor and control certified seeds, which has not been fully implemented, probably knowingly to manage the national food crisis [4].

Impacts on social movements

For some, Venezuelan defenders of AfroIndopeasant seeds cannot be considered a social movement, as they do not have homogeneous forms of organisation with clear agendas and programmes. However, there are diverse levels of articulation, exemplified by the National Peasant Seed Day, which continues to be celebrated to this day. As one activist puts it,

> we could say that we have some level of dispersion, because everyone returns to their territories to continue building their own spaces. But when it is necessary to advance transformations, there is a concrete articulation … the response is very fast [4].

New divisions emerge

The material conditions of the country have put a significant strain on social movement alliances. The 2015 seed law was in part developed as a transition pathway from agro-industry to agroecological and AfroIndopeasant *conuco*-based production. However, the current crisis makes previous agreements more tenuous. As described by seed activist Alma, some territories that were symbolic in the defence of AfroIndopeasant seeds and the *conuco* economy are now positioned as a key part of agribusiness [4].

For Alma, the material crisis of the Bolivarian Revolution revealed tensions between different movements that were not evident before. This is exemplified in the articulation that existed during the battles for the seed law with agrarian reform networks that were against large estates but still pursued industrialisation via mechanisation, hybrid seeds, etc. Alma sees in this case a disjunction between discourses and practices that affects the politics of the movement. While there may be commonalities in narratives, there is no coincidence in practices (e.g., agroecological and AfroIndopeasant *conuco*-based production). This might be related to 'the historical perspectives of the Left in general … with those notions of, basically, Modernity that still prevail in its agendas' [4].

Openings from COVID-19

With the implementation of COVID-19 restrictions, the 2020 National Peasant Seed Day had to move to a virtual space. Participating networks created a WhatsApp group, and the event was held in a decentralised and dispersed manner throughout the year, celebrated as each region saw fit [5]. This virtual space has become more important over the last few years, as the crisis led to shortages of petrol that collapsed public transport, and interruptions to public services like telephone lines became more frequent [4], [5]. Through this tool, interested parties are permanently connected; some 'groups keep in touch more, others don't, but everyone here knows who's who in the area of seeds and there's always interaction' [4].

WhatsApp chat participants are widely diverse and demonstrate the intercultural nature of the movement. There are representatives of peasant women, of urban agriculture, of Indigenous groups from the Amazon such as the Piaroa and Kumea peoples, or of education initiatives like *Todas las Manos a la Siembra* (All Hands to Sowing). There is even representation from all regions of the country, ethnicities, ages, and agricultural practices.

Participants are sometimes surprised to see who is in the chat group, as when the Minister of Urban Agriculture suddenly sent them a greeting [4]. This space has served several purposes:

> A lot of information of all kinds is shared, decisions of a more political nature [are made], but also particular techniques, calls for activities, exchange spaces. Over [the chat] we coordinated several of the seed exchanges that [have taken place], well, I think it fulfils an interesting function now in terms of the interconnectivity of that network [4].

Attacks to the law

Throughout 2016, a grassroots effort to inform the population about the new seed law was carried out in four regions of the country. Agribusiness attacks on the law intensified during this period, framing it as anti-scientific, anti-biotechnology, and biased towards local seed systems [3]. These attacks came especially from opposition[5] deputies who mocked the recognition of AfroIndopeasant seed practices and the *conuco* [2]. For some researchers, this resistance reveals a 'contentious politics of knowledge at play' [3, p. 668].

These attacks came as a surprise to some SGN participants, since their perspective is that the law did not affect industrial interests; it allows certified seeds to circulate, with institutions and procedures for their regulation, only prohibiting GMO seeds [5]. However, on the multiple occasions when the opposition has declared the collapse of the Bolivarian government and established 'ghost' governments or parallel assemblies, 'one of the first things they do is … put together a proposal for a seed law, … so it is clear that it is on their agenda' [5].

Implementation of the law

Despite government efforts, implementation of the 2015 seed law has been limited. The multidimensional crisis of the Bolivarian model has played a big role in this, as, for example, not enough technicians are being educated and the few who are working cannot travel to the regions due to the lack of petrol [4]. As mentioned in a previous section, some institutions have been created for the 'formal system' (e.g., RENASEM) but are not playing an active role in controlling certified seeds.

Some 'informal system' institutions contemplated in the law have not been created. A Popular Seed Council was proposed, but after some attempts to get it going regionally, it has not been established at national level. The networks created an online peasant seed information system called *el conuco*, but documentation of cases has been slow, as there are severe limitations to communication (e.g., blackouts or internet outages). As expressed by an activist, 'implementation has been lethargic due to the crisis' [4].

Implementation of the 2015 Seed Law has also seen some successes. The formation of the National Seed Commission and the development of the People's Seed Plan were achieved. The People's Seed Plan is 'an ongoing process of articulation among grassroots actors from across the country to coordinate efforts on the production and exchange of seeds' [3, p. 667]. Of particular interest to this research, the Scientific-Peasant Alliance is being implemented by the Corporation for Scientific and Technological Development (CODECYT).

This initiative joins technoscientific knowledge and popular knowledge(s) to develop agroecological alternatives for food production. The alliance has a place-based approach, whereby the *conuco* is recognised as a space of learning, local capacities are acknowledged, and family is understood as the centre of productive and cultural processes. The alliance includes over 5,000 peasant families organised in seeding groups across Venezuela, with 62 per cent of participants being women or young people [31]. A guardian manifested their faith in the advancements of the alliance in the potato sector:

> There are some very interesting processes of creating technical spaces for improvement, germ-plasm banks, etc. managed by social groups. In other words, it's not that [they] created an intermediary to manage [it] for people; the technicians are sons [and] daughters of peasants trained for it [5].

Now nearing ten years since its foundation, the Scientific-Peasant Alliance has contributed to the recovery of native varieties, seed diversification, and advancing an agroecological model at national scale (e.g., by substituting agrochemicals). The alliance has consolidated socio-productive networks throughout the country, integrating families, community councils, cooperatives, and research centres. Processes of knowledge production occur collaboratively as new agricultural inputs and techniques are tested directly on plots and with farmers. Producing and disseminating knowledge through *diálogo de saberes* decolonises science by revaluing previously invisible ancestral techniques and using a producer-to-producer methodology where the most experienced farmers become local trainers. Accordingly, the Venezuelan government frames the alliance as a mechanism for decolonising agriculture [32].

As alternative institutions consolidate, SGNs continue their work articulating stakeholders for the defence of native seeds in the face of a multidimensional crisis, which has reconstituted agribusiness in the country. As described by seed activist Alma,

> connecting all these initiatives will reconfigure the Venezuelan movement for the defence of seeds. If you had interviewed me earlier, I would have given you a totally different answer, but I have come from three meetings that have made me realise that the people are there, they are active, and they are mobilising [4].

Conclusion

So, what has contributed to change? This book sees Venezuelan seed system transformation as a nested process of change starting from the 1999 Bolivarian constitution (see Figure 5.1). This and subsequent legislation were framed as steps towards eco-socialism, including a push for agroecology and food sovereignty. In fact, the whole Bolivarian Revolution is framed as a transition from capitalist extractivism to a participatory socialism where humans and nature coexist harmoniously.

In parallel to the Bolivarian Revolution, seed guardian networks and other adjacent social movements consolidated through loose narrative connections and an interest in protecting and encouraging AfroIndopeasant food practices. These actors took advantage of novel participatory democracy instruments to be involved in the development of the 2015 seed law. Eventually, a process of popular debate led to the production of the most progressive seed law in Latin America, a law that recognises and promotes the formal and informal(ised) seed system. This process can be considered as the opening of a translational contact zone, as it required diverse actors to see beyond the epistemic, legal, and scientific abyssal lines.

Implementation of the 2015 seed law has been sluggish due to the multidimensional crisis in Venezuela and resistance from various food system actors. More research is needed to understand emerging Transmodern institutions and their impact. However, a Transmodern seed has been planted in Venezuelan food systems as more spaces for epistemic justice and horizontal human–nature relations are opened.

It is important to emphasise that the Bolivarian model has brought suffering to the Venezuelan people. State institutions have weakened, hunger and economic hardship are spreading, and millions of Venezuelans have left the country. Although there is no consensus on the causes, this work recognises the failings of the Bolivarian model while also acknowledging that there are external forces at play (e.g., economic sanctions by rich nations). In January of 2026, the government of the United States of America captured President Nicolas Maduro after a brief military incursion into Venezuelan territory. The future of the Bolivarian eco-socialist model seems increasingly precarious. However, as this chapter has evidenced, seed guardians will keep advancing seed sovereignty against all odds.

Notes

1 *Conuco* refers to a traditional agroforestry system based on small parcels generally connected to a family home, polycultures, and family agriculture.
2 *Red de Productores Integrales del Páramo.*
3 *Pagamento* is the act of making ritualistic offerings to land, food, spirits, etc. for good crops.
4 The Spanish phrase *maestres pueblo* (popular teachers or knowers) refers to local knowledge-holders, particularly those community experts with experience of local seed varieties. In the legislation, they are referred to in the masculine as *maestros pueblo*, but here the author chooses to use the genderless *maestres.*
5 Opposition came from the 'Democratic Unity Roundtable' party (MUD, from its initials in Spanish).

References

[1] C. Rory, *Comandante: HugoChávez's Venezuela*. New York: Penguin Books, 2014.
[2] A. Felicien, C. M. Schiavoni, and L. Romero, 'Food politics in a time of crisis: Corporate power vs. popular power in the shifting relations of state, society and capital in Venezuela's food system', Conference paper no. 9. ERPI 2018 International Conference Authoritarian Populism and the Rural World, The Hague, International Institute of Social Studies (ISS), 2018.
[3] A. Felicien *et al.*, 'Exploring the "grey areas" of state-society interaction in food sovereignty construction: The battle for Venezuela's seed law', *Journal of Peasant Studies*, vol. 47, no. 4, pp. 648–673, 2020, doi:10.1080/03066150.2018.1525363.
[4] NET007, 'private communication', Oct.2021.
[5] NET008, 'private communication', Oct.2021.
[6] A. Felicien, 'Historia de las leyes de semillas y la modernidad en la agricultura', in *Semillas del pueblo: Luchas y resistencias para el resguardo y reproducción de la vida*, 1st ed., M. Pérez, A. Felicien, S. Saturno, E. Ochoa, C. V. L. de Transgénicos, L. Buitrago Arévalo, L. Romero, B. Torres, B. Silva, J. Toro, and X. Gonzáles Broquen, Eds. Caracas: Editorial La Estrella Roja, 2016, pp. 21–49.
[7] Venezuela. Asamblea Nacional Constituyente. (1999). *Constitución de la república Bolivariana de Venezuela*. Caracas: Asamblea Nacional Constituyente.
[8] E. Minaya and K. Vyas, 'Protesters in Venezuela press government', *The Wall Street Journal*, 23 Feb.2014.
[9] L. Buitrago Arévalo, 'La semilla como elemento articulador de los movimientos sociales', in *Semillas del pueblo: Luchas y resistencias para el resguardo y reproducción de la vida*, 1st ed., M. Pérez, A. Felicien, S. Saturno, E. Ochoa, C. V. L. de Transgénicos, L. Buitrago Arévalo, L. Romero, B. Torres, B. Silva, J. Toro, and X. Gonzáles Broquen, Eds. Caracas: Editorial La Estrella Roja, 2016, pp. 73–86.
[10] J. E. Rodríguez Rojas, 'Contra-reforma, inestabilidad macroeconómica y autonomía del sistema agroalimentario Venezolano durante el periodo 1989–2006', *Agroalimentaria*, vol. 15, no. 29, pp. 39–53, 2009.
[11] A. Felicien, 'Resistencias y disputas alrededor de las semillas en Venezuela', in *Semillas y crisis alimentaria en Venezuela*. Quito: Fundación Rosa Luxemburg Oficina Región Andina, 2021.
[12] E. Ochoa, 'Leguminosas en resistencia: La relación histórica conuco-leguminosas y su relevancia cultural en Venezuela', in *Semillas del pueblo: Luchas y resistencias para el resguardo y reproducción de la vida*, 1st ed., M. Pérez, A. Felicien, S. Saturno, E.

Ochoa, C. V. L. de Transgénicos, L. Buitrago Arévalo, L. Romero, B. Torres, B. Silva, J. Toro, and X. Gonzáles Broquen, Eds. Caracas: Editorial La Estrella Roja, 2016, pp. 171–189.

[13] F. Miranda and L. Urdaneta, 'Evolución del marco legal de la producción de semilla en Venezuela', in *Producción de Semillas en Venezuela*, H. Laurentín, Ed. Maracaibo: Ediciones Astro Data, 2020, pp. 15–43.

[14] Venezuela. Asamblea Nacional de la República Bolivariana de Venezuela. (2002). *Ley de tierras y desarrollo agrario.* Accessed: 10 Dec. 2025. [Online]. Available: https://faolex.fao.org/docs/pdf/ven28661.pdf.

[15] Venezuela. Asamblea Nacional de la República Bolivariana de Venezuela. (2008). *Decreto 6.129 con rango, valor y fuerza de ley de salud agrícola integral.* Accessed: 10 Dec. 2025. [Online]. Available: https://drive.google.com/file/d/1t46vKJawxKDvR7tXqoys2vTLH3cSknsx/view.

[16] Venezuela. Asamblea Nacional de la República Bolivariana de Venezuela. (2008). *Ley de gestión de la diversidad biológica.* Accessed: 10 Dec. 2025. [Online]. Available: https://www.asambleanacional.gob.ve/storage/documentos/leyes/ley-de-ges-20220131154108.pdf.

[17] Venezuela. Asamblea Nacional de la República Bolivariana de Venezuela. (2015). *Decreto No 6.071, con rango, valor y fuerza de ley orgánica de seguridad y soberanía agroalimentaria.* Accessed: 10 Dec. 2025. [Online]. Available: https://www.asambleanacional.gob.ve/storage/documentos/leyes/decreto-n0-20220201144829.pdf.

[18] Venezuela. Ministerio del Poder Popular para la Educación. (2009). *Resolución Ministerial No 024.* Accessed: 10 Dec. 2025. [Online]. Available: https://www.scribd.com/document/206854641/Resolucion-024.

[19] S. Salcedo and L. Guzmán, Eds., *Agricultura familiar en América Latina y el Caribe: Recomendaciones de política*, 2014. Accessed: 28 Jan. 2026. [Online]. Available: https://www.fao.org/4/i3788s/i3788s.pdf.

[20] Instituto IAEPHCH, 'Aló presidente No 189 (18/04/2004)', 2004. Instituto Hugo Chávez. Accessed: 10 Feb. 2026. [Online]. Available: https://chavezporsiempreiaepch.blogspot.com/p/18042004.html.

[21] A. Gutiérrez, 'El sistema alimentario Venezolano (SAV): Evolución reciente, balance y perspectivas', *Agroalimentaria*, pp. 19–60, 2015. Available: www.proquest.com/magazines/el-sistema-alimentario-venezolano-sav-evolución/docview/1700707573/se-2.

[22] Ministerio del Poder Popular para la Agricultura y Tierras, 'Encuentro Internacional de Guardianes de Semillas', Prensa MinAgricultura, 2012. Accessed: 10 Dec. 2025. [Online]. Available: https://prensamat.blogspot.com/2012/10/encuentro-internacional-de-guardianes.html?q=gaudy.

[23] GRAIN, 'De guardianes, ferias y casas de semillas', 2017. Accessed: 12 Jun. 2022. [Online]. Available: https://www.grain.org/media/W1siZiIsIjIwMTcvMDgvMDgvMTBfMzZfMzNfOTY2XzAyRGVfdW5fdmlzdGF6b195X211Y2hhc19hcmlzdGFzLnBkZiJdXQ.

[24] Campaña Venezuela Libre de Transgénicos, 'Cronología: Proceso de construcción colectiva de la ley de semillas en Venezuela', in *Semillas del pueblo: Luchas y resistencias para el resguardo y reproducción de la vida*, 1st ed., M. Pérez, A. Felicien, S. Saturno, E. Ochoa, C. V. L. de Transgénicos, L. Buitrago Arévalo, L. Romero, B. Torres, B. Silva, J. Toro, and X. Gonzáles Broquen, Eds. Caracas: Editorial La Estrella Roja, 2016, pp. 51 71.

[25] Semillero Socialista de Monte Carmelo, 'I° Encuentro Internacional de Guardianes de Semillas', 2012. Accessed: 3 Feb. 2026. [Online]. Available: https://cdn.biodiversidadla.org/content/download/95255/631520/file/Encuentro+Internacional+de+Guardianes+de+Semillas+MONTE+CARMELO.pdf.
[26] Venezuela. Asamblea Nacional de la República Bolivariana de Venezuela. (2015). *Plan de la patria: Segundo plan de desarrollo económico y social de la nación 2013–2019*, Agroalimentaria, pp. 19–60. Available: www.proquest.com/magazines/el-sistema-alimentario-venezolano-sav-evolución/docview/1700707573/se-2.
[27] Congreso Venezolano de Diversidad Biológica, 'Declaración del VI Congreso Venezolano de Diversidad Biológica', n.d. Biodiversidadla. Accessed: 10 Feb. 2026. [Online]. Available: www.biodiversidadla.org/Documentos/Declaracion_del_VI_Congreso_Venezolano_de_Diversidad_Biologica.
[28] M. Pérez, 'Experiencias de reconocimiento, rescate y producción de semillas campesinas, indígenas y afrodescendientes desde el poder popular en Venezuela', in *Semillas del pueblo: Luchas y resistencias para el resguardo y reproducción de la vida*, 1st ed., M. Pérez, A. Felicien, S. Saturno, E. Ochoa, C. V. L. de Transgénicos, L. Buitrago Arévalo, L. Romero, B. Torres, B. Silva, J. Toro, and X. Gonzáles Broquen, Eds. Caracas: Editorial La Estrella Roja, 2016, pp. 87–152.
[29] Venezuela. Asamblea Nacional de la República Bolivariana de Venezuela. (2015). *Ley de Semillas*. Accessed: 10 Dec. 2025. [Online]. Available: www.asambleanacional.gob.ve/storage/documentos/leyes/ley-de-semillas-20211025164906.pdf.
[30] The UN Refugee Agency, 'Venezuela situation', n.d. UNCHR. Accessed: 10 Feb. 2026. [Online]. Available: www.unhcr.org/emergencies/venezuela-situation.
[31] Ministerio del Poder Popular para la Ciencia y Tecnología, 'Alianza Científico-Campesina: Producción, ciencia y trabajo para la preservación de la semilla Venezolana'. Accessed: 10 Dec. 2025. [Online]. Available: https://mincyt.gob.ve/alianza-cientifico-campesina-produccion-ciencia-y-trabajo-preservacion-semilla-venezolana.
[32] Ministerio del Poder Popular para Ciencia y Tecnología, *Alianza Científico-Campesina Ciencia: saberes ancestrales y Buen Vivir al servicio de la soberanía alimentaria de Venezuela*, 2025. [Online]. Available: https://mincyt.gob.ve/wp-content/uploads/2025/08/LIBRO-alianza-cientifico-campesina-digital.pdf.

6 Ecuadorian seed system transformations

The early 1990s marked the 500-year anniversary of the Iberian invasion of America. This momentous occasion stirred Indigenous social movements all over Latin America, but particularly in Ecuador, where they became a political force to be reckoned with. Two important Indigenous-led events occurred during those early years of the 1990s: the formation of the Confederation of Indigenous Nationalities of Ecuador (*Confederación de Nacionalidades Indígenas del Ecuador*, CONAIE), and the emergence of a Quichua neologism in the Amazon – *Sumak Kawsay*; an alternative way of being in harmony with oneself, community, and nature.

Following historical trends, these peripheral emergences would have been kept on the sidelines, but in this case they went on to transform Ecuadorian politics two decades later. *Sumak Kawsay*, later translated into Spanish as *Buen Vivir*, can be considered as an umbrella term with a purposefully vague definition that encourages diverse alternative ways of relating to self, society, and nature. The term emerged from Indigenous lifeworlds in response to the global ecological crisis, as Amazonian communities were confronted by 'Global North' definitions of sustainability.

However, *Buen Vivir* is not purely an Indigenous concept but a fruit of transcultural interactions between Indigenous communities, development agencies, and an Indigenous and mestizo 'indigenist intelligentsia'. The concept remained on the peripheries of intellectual production until it crossed into the mainstream through academic publications, becoming a central theme in Ecuadorian social struggles in the early 2000s.

The beginning of the 21st century was a critical juncture for Ecuador. The country was a young democracy, with its first ever election held in 1979 after decades of military and populist rule [1]. From 1979 to 2006, Ecuador had a dozen presidents, most of whom faced staunch resistance from rural and Indigenous social movements as neoliberal reforms were implemented and the economy was dollarised.

In fact, the 1996–2006 period saw three presidents ousted by social movements: Abdalá Bucaram (1997), Jamil Mahuad (2000), and Lucio Gutierrez (2005) [2], [3]. The main opposition during this period of political instability was the CONAIE [4], which went on to create the Indigenous political party Pachakutik in the early

DOI: 10.4324/9781003715092-9

2000s, becoming a key player in Ecuadorian politics. At this critical juncture, *Sumak Kawsay/Buen Vivir* was mobilised politically as an alternative pathway.

After more than a decade of political crisis, Rafael Correa ran for office in 2006, without political party affiliation as a rejection of the establishment. His platform promised the launch of a 'citizen's revolution' that would create a new constitution to guide the country into the new millennium. Social movements representing marginalised sectors of Ecuadorian society supported Correa, putting him onto the highest seat of power in 2007 with the mandate of organising a constitutional assembly. After more than a year of national discussions, a new constitution was approved using *Buen Vivir* as a guiding principle. It enshrined the rights of nature, included the right to food sovereignty, and cemented institutions for participatory democracy.

The 2008 constitution led to the creation of a development plan based on *Buen Vivir* principles [5], the Organic Law of Food Sovereignty (2010), the Organic Law of Rural Lands and Ancestral Territories (2016), and the creation of the Plurinational and Intercultural Conference on Food Sovereignty (*Conferencia Plurinacional e Intercultural de Soberanía Alimentaria*, COPISA). This organisation began drafting a new seed law in 2014, which was finally approved in 2018 as the Organic Law of Agrobiodiversity, Seeds, and Promotion of Agriculture.

The government of Rafael Correa and his citizen's revolution (2007–2017) have been widely criticised for betraying social movements and abandoning *Buen Vivir* principles to implement a neo-extractivist agenda. However, it cannot be denied that this period of Ecuadorian history was a fertile space for the emergence of novel legal frameworks and institutions.

This chapter presents the transformation leading to the 2018 seed law beginning with the emergence of *Sumak Kawsay* in the early 1990s (See Figure 6.1). The openings afforded by *Buen Vivir* and its irruption in the foundational narrative of the Ecuadorian state in 2008, allowed for broad inclusion of non-Modern practices. This process was moved forward by AfroIndopeasant social movements at multiple scales with an active participation of seed guardian networks. Ecuador has been touted as having the most advanced legal and policy framework for food sovereignty in the world [4]. In this transformation process, Indigenous visions of alternative futures open spaces for non-Modern food practices to be recognised in policy, and to be protected and encouraged by state actors.

Sumak Kawsay (*Buen Vivir*)

The emergence of *Buen Vivir* is positioned in this text as the first step for the consolidation of a Transmodern seed system in Ecuador. This alternative vision of wellbeing spread throughout the nation and was critical for the formation of coalitions and the creation of platforms to transform politics. This concept would guide the 2008 constitution, which kickstarted a legislative boom that opened new spaces for food sovereignty, native seeds, and related care practices.

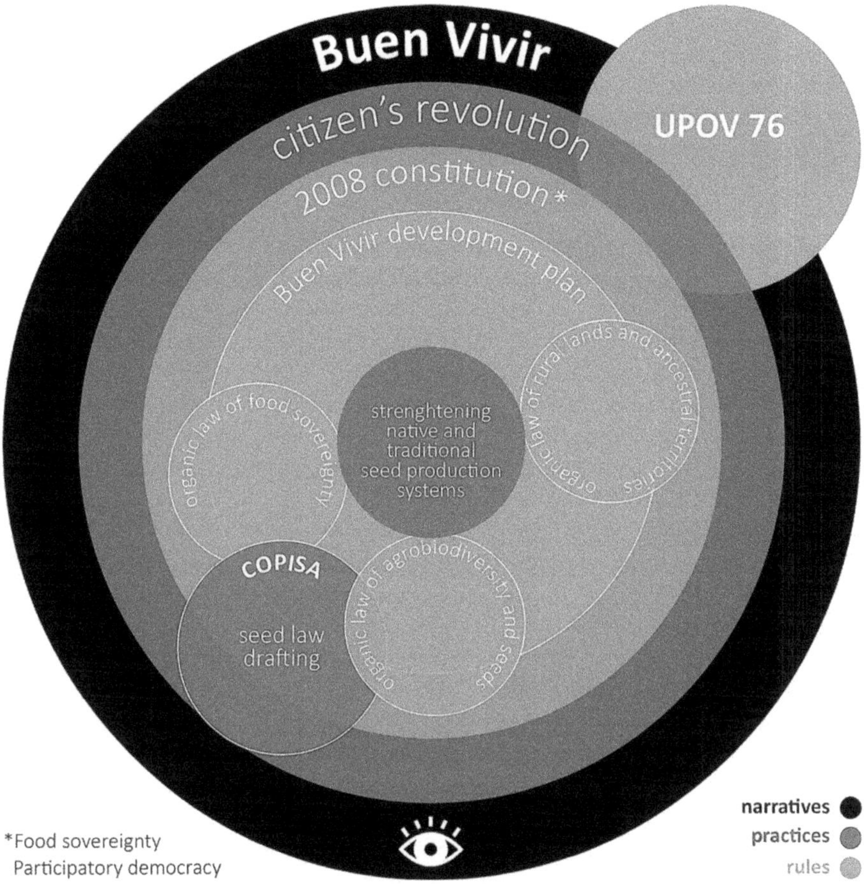

Figure 6.1 A visualisation of Ecuadorian seed system transformations as a series of nested fields emerging from previous structuration

Buen Vivir*: Discourse and platform*

Buen Vivir is often defined as a way of living in harmony with oneself, community, society, and nature [6], [7], [8], [9], [10]; in short, a plurality of good coexistences [11]. More generally, *Buen Vivir* can be understood as a set of ideas that are simultaneously 'a radical critique of Development ... and a set of alternative proposals where well-being is only possible in socially and ecologically expanded communities' [12, p. 6].

Buen Vivir is a discursive innovation [13], [14] that does not pursue a return to an idealised past [12], [15], [16], [17] but looks to the future while conceptualising and building alternative utopias [7], [12]. Accordingly, *Buen Vivir* can be understood variously as: a 'process of collective social innovation' [18, p. 152]; a 'conceptual and affective platform, another way to understand the world, or even another cosmovision' [12, p. 9]; an epistemic alternative and possible 'line of flight'[1] from the dominant economic paradigm [19]; or a 'rhetorical-ideational resource that breaks with dominant language and mental models, giving it a subversive political and cultural character facing the advance of a totalising and globalising Eurocentric modernity' [12, p. 15].

Buen Vivir emerges from Indigenous responses to the multiple crises of our epoch, consolidating in the Amazonian Quichua neologism *Sumak Kawsay* [20]. *Buen Vivir* has Indigenous origins, but it is an assemblage of Indigenous notions and Western critiques of Western modernity, including those with Marxist, feminist, and ecological influences [12], [19], [21]. As such, it can be understood as glocal [12] and dialogical [14], [16]. Indeed, *Buen Vivir* is the result of interactions by multiple actors at many levels: Indigenous communities, intellectuals and activists, diverse academics, multiple social movements, leftist politicians, and NGOs.

Buen Vivir has been dubbed a 'collage construct' [7, p. 39], relevant not because of 'its pure expressions but by the interfaces it proposes' [9, p. 192]. This refers not only to the integration of diverse Indigenous and Western(ised) concepts, but also to its movement and feedback loops from bottom-up to top-down [7], [9].

Buen Vivir has been referenced in the literature as a discourse, a discursive articulation, or discursive apparatus alternative to traditional notions of Western development [12], [14], [21]. It is a concept under constant construction with diverse interpretations [12], [16]. Accordingly, *Buen Vivir* has been referred to as a polysemic notion [21], an ambiguous concept, a 'semantic field in motion' [9], and a 'polychrome and ambivalent discourse' [12, p. 15].

Critics contend this ambiguity raises the risk of *Buen Vivir* being interpreted as an empty term simultaneously meaning everything and nothing [7], [9]. Some authors see *Buen Vivir* as a 'neologism built ad hoc' [21, p. 79], since there is little to no ethnographic evidence of Indigenous use of the term before the 21st century [19], [21]. Some more critical voices call it an 'invented tradition' produced by an idealised and ahistorical vision of Andean communities [22, p. 64].

While these critiques are valid, the neologism can also be interpreted as the revaluing of a previously marginalised design for life [11] with the potential to face the multiple global crises of our current epoch.

Discursive spread of Buen Vivir

The emergence of *Buen Vivir* has its roots in the historical resistance of Indigenous peoples of Latin America to colonisation and Westernisation [11]. Despite this being a 500-year struggle, multiple forces converged in the second half of the 20th century to set the stage for *Buen Vivir* flourishing. It began in the 1960s with the emergence of environmentalism, the consolidation of multiculturalism, and a renewed Western interest in the 'good life' [12]. It continued with the start of neoliberal implementation in the 1970s across Latin America (often enforced by dictatorships) [23], and then the end of the Cold War and a slow return to democracy in the region that prompted a revitalisation of civil society, collective action, and a revaluation of the political left [12].

Parallel to these trends, international cooperation agencies started disseminating concepts like 'sustainable development' and 'development with identity' amongst Indigenous peoples in the late 1980s. These moves were heavily resisted by Indigenous and mestizo Amazonian intellectuals who sought to pursue alternatives to Development instead of alternative Development [7]. Significantly, by the end of the decade, 24 nations ratified the International Labour Organization's Indigenous and Tribal Peoples Convention, in which Central and South American countries made up an overwhelming majority of signatories.

The early 1990s brought a wave of constitutional reforms to several members of the Andean Community of Nations,[2] processes that reformulated monolithic and homogenising nation-states into plurinational, multi-ethnic, and intercultural countries [12], [19]. These reforms aligned with the 500-year anniversary of the invasion of America in 1992, an event that spawned numerous mobilisations against the colonial celebration and led to a strengthening of Indigenous social movements [11], [12], [24]. Crucially, the Amazonian concept of *Sumak Kawsay* as an Indigenous alternative to development also made its first written appearance in 1992 [7].

The Indigenous political resurfacings of the 1990s[3] converged with an increasing consciousness of humanity's multiple systemic crises and concerns about the failed promises of neoliberalism. This pushed citizen movements to develop stronger critiques of market policies, and to eventually question and seek alternatives to neoliberal development [12], [14]. Accordingly, this decade saw the consolidation of the decolonial and post-Development academic fields [24], [25], [26]. As described previously, the emergence of *Buen Vivir* can be interpreted as a historically fortuitous glocal convergence of multiple struggles at various scales [12].

Buen Vivir emerged with full force in the first decade of the 21st century [19] along with the continued resistance of Indigenous, Afro-Latino, peasant, and other social movements against globalisation and the privatisation of natural resources [12]. In Latin America, the new millennium brought a slew of left-wing governments that (unlike many of their predecessors) had access to resources due to high commodity prices [12]. This phenomenon has been dubbed the 'pink tide', and included Argentina, Bolivia, Brazil, Chile, Ecuador, El Salvador, Honduras, Nicaragua, Panama, Paraguay, Peru, Uruguay, and Venezuela.

The Ecuadorian pink tide was mainly borne along by Indigenous cultural discourses (*Sumak Kawsay*), enabling the development of *Buen Vivir* experiments at national level. It was later in the decade that the discursive innovation of *Sumak Kawsay* truly crystalised by being incorporated into the Ecuadorian constitutional assembly [11], [12], [21]. The approval of a *Buen Vivir*-influenced constitution in 2008 led to the creation of a *Buen Vivir* development plan for 2009–2013.

Ecuador's constitutional process coincided with the United Nations' Declaration on the Rights of Indigenous Peoples (2007) and the economic crisis of 2008, which produced further global questionings of the neoliberal order and a generalised loss of faith in the political class [12], [21]. One would think the stage was set for a state-supported move to a biocentric post-capitalist society. However, national implementations of *Buen Vivir* have not been completely successful and are often criticised for being more reformist than transformative [17], [21], [27].

For some, state capture of *Buen Vivir* led to a loss of its decolonial potential and 'cutting-edge positionality as a subaltern philosophy and political strategy' [23, p. 425]. The Ecuadorian government has been accused of abandoning, and even betraying, its previous alliances by becoming merely progressive instead of embracing plurality [12]. Some frame governmental *Buen Vivir* projects as a continuation of the same old developmentalist model using the discourse as powerful new branding [17], [21], [28].

Despite all its failings at a state level, *Buen Vivir* is and always has been present in the lived praxis of countless communities throughout the region [19]. Additionally, *Buen Vivir* is still informing processes at multiple levels and in multiple geographies, 'shaping discourse, political projects, and social, cultural, and economic practice' [11, para. 16].

Citizen revolution (2007–2017)

In early 2005, the Ecuadorian congress removed President Lucio Gutierrez from power in what has been referred to as a coup d'état [29]. The then vice-president assumed power until early 2007, when Rafael Correa was sworn in. Correa ran for president as a political establishment outsider, promising that he would convene a Constituent Assembly that would 'collect demands for the refoundation of the republic and … adopt a development model different from neoliberalism' [29, p. 38].

This anti-neoliberal narrative differentiated Correa from an opposition stuck in the old ways and without a clear vision, which eventually led Correa to garner the support of social movements [30]. Correa mobilised narratives of a citizen's revolution and 21st century socialism to communicate his political project. At the time, some criticised these narratives as lacking any clear definition, being used only as slogans to justify political action [31]. The following sections explore the outlines of these narratives and the legislation produced by Correa's citizen's revolution.

21st century socialism in Ecuador

21st century socialism was a narrative mobilised throughout the region as the pink tide expanded. As Ellner explains it:

> Leaders … created a new narrative of nationhood that challenges long-held assumptions and previous representations of culture, history, race, gender, citizenship, and identity. Thus, the new political movements offer

> an alternative reading of the past that challenges the conventional wisdom that legitimised the old order … Re-envisioning the past serves to incorporate previously marginalised peoples, including the Indigenous, Afro-descendent, peasants, women, and workers who historically struggled to change social conditions [30, p. 107].

Some of the contours of 21st century socialism have been partially defined in parallel to its implementation throughout Latin America. Ellner [30] sees the distinguishing features of this 'new left in power' as:

- A defence of radical democracy (social incorporation and direct participation).
- A pursuit of alternatives to capitalism, mainly socialism adapted to the concrete realities of Latin America.[4]
- A rejection of working-class vanguardism, or at least a decentring of the industrial proletariat, while seeking the inclusion of the precariat, informal economies, religious communities, AfroIndopeasants, and women.
- A focus on social wellbeing, fraternity, and solidarity while rejecting capitalist individualism.[5]
- A push for a multi-polar world, while still operating within globalised institutions; self-identifying as anti-capitalist (clashing with the agendas of Washington, DC) and strengthening relations in the region.[6]
- A personification of popular will through historical symbols.[7]

Neoliberal Ecuador was marked by an absentee state in social and economic life. The citizen's revolution pursued a return and expansion of the state, through increased public spending, and the growth of ministries and other government institutions [2]. This period drew criticisms from all sides. For some, the citizen's revolution ended in a state-guided neo-developmentalist plan mostly inspired by classical social-democratic templates rather than by an alternative paradigm [4], [18].

Using food systems as an example, distribution of funds was grossly unequal between regions [32], investment mainly benefited medium- and large-scale producers, conventional agriculture was preferred to agroecology, and presidential declarations reinforced narratives of small-scale production as inefficient [2]. Other critics stated that favouring inclusion instead of production targets translated into 'belittling technical competence and efficiency' [30, p. 110]. Inclusion itself has been criticised, as social movement capacity to mobilise and organise has diminished as leaders were co-opted into government and NGO bureaucracies [2].

In short, the citizen's revolution did not become what was initially envisioned. The resilience of pre-existing economic powers, the weakening of social movements, clientelism, a hyper-presidential structure, and the delegitimisation/criminalisation of government critics eclipsed its broader purpose [18].

Reconstituting Ecuador

In 2006, Rafael Correa positioned himself as an outsider candidate to the political establishment. As such, social movements carried him to the presidency to fulfil his promise of pursuing alternatives. In his early days as president, he did not have institutional support to carry out plans to produce a new national constitution [31]. This led to a convoluted process that set the stage for the citizen's revolution, with all its possibilities and eventual broken promises.

The constitutional process in Ecuador is the key event that opens new pathways for the establishment of legal frameworks and institutions inclusive of food sovereignty and native seed care. As it will be elaborated below, it had profound impacts in national seed system transformation.

Constitutional assembly (2007–2008)

As soon as Rafael Correa was sworn in as president, he called for a referendum to create a new constitution. This was not a seamless process, as conflicts emerged between the executive and the Ecuadorian congress, leading to the removal of 57 deputies opposed to the constitutional process. After this legislative crisis, a constitutional referendum was carried out in early 2007 with upwards of 80 per cent of voters mandating the creation of a constituent assembly.

In September of 2007, an election was held to choose 130 assembly members representing diverse sectors of Ecuadorian society. Through this election, a loose coalition aligned with Rafael Correa (Alianza PAIS) won more than 70 per cent of available seats. This overwhelming majority was the product of a divided opposition contrasted with a plurality of actors coming together in the Alianza PAIS coalition: middle-class dissidents from Quito, urban middle-class actors loosely affiliated with CONAIE, and local-level leaders from peri-urban organisations, among others [29]. The resounding win by Alianza PAIS was interpreted as a defeat of traditional powers and meant that the constitutional assembly would become 'the axis for power struggles in Ecuador' [32, p. 10].

Andrade distinguishes two moments in the constitutional assembly: an installation phase where Alianza PAIS created the rules of the game, and a decisional phase where the definitive texts where decided. In the installation phase, working roundtables were organised around three axes: '[d]eepening of democracy towards participatory democracy, an ecologically sound development model, and the notion of complementarity between the state, nature and society' [29, p. 39]. Roundtables were organised as follows:

- Fundamental Rights and Constitutional Guarantees.
- Organisation, Social, and Citizen Participation.
- State Structure and Institutions.

- Territorial Planning and Power Assignment.
- Natural Resources and Biodiversity.
- Work, Production, and Social Inclusion.
- Development Regime.
- Justice and the Fight against Corruption.
- Sovereignty, International Relations, and Integration.
- Legislation and Oversight.

Going deep into the discussions of each working roundtable exceeds the purpose of this book, as actors and issues discussed were widely diverse. However, these can be broadly understood as a bricolage between neo-constitutionalist, participationist, and presidentialist perspectives [29].

Despite the elected representatives being tasked with moving the constitutional process forward, this was also a participatory and decentralised effort [3]. Intellectuals, and Ecuadorian and foreign NGOs were very active in the process, writing and disseminating opinion articles, as well as giving speeches to the assembly [1]. Social organisations also developed their own proposals,[8] formed their own parallel working roundtables, and collaborated directly with legislators [3].

As part of this participative move, an Indigenous and peasant coalition formed a parallel agrarian roundtable to create proposals based on food sovereignty framings focusing on domestic distribution and markets [33]. The Ecuadorian seed guardian network took advantage of these participatory spaces to be advisors to the assembly and help draft the new constitution [34], a drafting process that some authors understand as the 'critical political opening for food sovereignty in Ecuador' [35, p. 226].

For some, the Ecuadorian constitutional process attempted to refound the state [4]. It was a historical opportunity to 'decolonise the country's political structures' [36, p. 48], as it sought to inspire new alternative public practices and socioeconomic models, building an understanding of the world from those who have been historically marginalised [11]. Some went as far as claiming that it would bring forth a biocentric turn [37].

These optimistic framings are frequently criticised, since practical implementations by the state have been far from alternative or post-capitalist, and unable to create a rupture from Eurocentric, colonial, and capitalist frameworks [22]. The process itself was also criticised as an example of plebiscitary democracy, where the executive power frames issues from its own agenda, presenting the constitutional referendum as a zero-sum game [27]. The validity of these critiques is unquestionable. However, the constitutional text does diverge greatly from established frames, as it includes *Buen Vivir*, direct democracy, the rights of nature, and food sovereignty.

2008 Constitution

The constitutional text co-produced in 2008 reflects the inclusion of multiple marginalised actors in Ecuadorian society. This section focuses on items that refer to *Buen Vivir*, plurality, nature, and food sovereignty, as they provide the foundations for national seed system transformations. In its preamble, millenary roots and women are explicitly recognised and Pachamama[9] is celebrated. It appeals to the wisdom of all cultures present in the national territory and connects to past liberation struggles against all forms of domination and colonialism. The preamble also frames Ecuador as an intercultural and plurinational state, emphasising the legacy and present of ancestral peoples.

Crucially, the text is positioned as a decision to create a new way to coexist 'in diversity and harmony with nature, to achieve Buen Vivir, the sumak kawsay' [38, p. 15]. The development model proposed in the constitution is geared towards the attainment of *Buen Vivir*, and citizens are expected to act in a framework of interculturality, respect of diversities, and harmonic coexistence with nature [38].

Regarding rights, the constitution positions nature as a subject of rights and establishes several *Buen Vivir* rights. These include the right to water and access to food with a commitment of promoting food sovereignty, along with the right to a healthy environment, and promoting clean technologies for energy production. Also included is the right to an education that is 'participatory, ... intercultural, ... inclusive and diverse, of quality and warmth;[10] promoting gender equality, justice, solidarity and peace' [38, p. 27].

The rights of Indigenous, Afro-Ecuadorian, and Montubio[11] peoples have special consideration, seeking to preserve and promote their practices of biodiversity management, develop their collective knowledge (framed as ancestral sciences and technologies), as well as agrobiodiversity. The text also includes an explicit prohibition of all forms of appropriation of these knowledge(s), innovations, and practices.

Expanding the spectrum of rights, Chapter 7, Article 71 of the constitution states that:

> Nature, or *Pacha Mama*, where life is reproduced and occurs, has the right to complete respect for its existence and for the maintenance and regeneration of its life cycles, structure, functions, and evolutionary processes. All persons, communities, peoples, and nations can call upon public authorities to enforce the rights of nature. To enforce and interpret these rights, the principles set forth in the Constitution shall be observed as appropriate. The state shall give incentives to individuals and legal entities and to communities to protect nature and to promote respect for all the elements comprising an ecosystem [38, p. 52].

It is important to point out that both right- and left-wing politicians have condemned the inclusion of legal rights of nature as an infantile strategy of indigenist politics [27], [39]. This might be interpreted as evidence that, despite critiques to the constitutional process and its impacts, it clearly is proposing alternatives.

Regarding rights to participation, all Ecuadorians can put forward normative projects as popular initiatives. The constitution contemplates popular control of state institutions, society, and its representatives, and establishes the mechanisms of representative, direct, and communitarian democracy for this purpose. Title IV, Chapter 1, Section 4 defines direct democracy, establishing 'normative popular initiatives' as a mechanism for citizens to create, reform, or strike down laws if they have support from 0.25 per cent of people registered to vote. Additionally, Title VI, Chapter 2 proposes a national decentralised participatory planning system to establish development guidelines.

Of special interest to this book is the inclusion of food sovereignty as a strategic objective and a state obligation. This constitution was one of the first in the world to explicitly recognise food sovereignty as a goal [33]. This was a very controversial topic, since some provisions contained highly contested issues (mainly the introduction of GMOs) [3]. For some authors, this effort to include food sovereignty was diluted, as it compromised to include many visions of the concept, leading to residual pro-poor projects 'rather than a broader transformative pathway with comprehensive agrarian reform' [40, p. 1196].

Some important state responsibilities related to food sovereignty are: (i) promoting the production and transformation of agri-food and fishing production units at the small and medium scale, community scale, and belonging to the social and solidarity economy; (ii) strengthening diversification and introducing ecological and organic technologies into agricultural production; (iii) promoting the preservation and recovery of agrobiodiversity and associated ancestral knowledge, as well as the use, conservation, and free exchange of seeds; and (iv) protecting the population from consuming contaminated foods, or foods that put their health at risk, or whose effects are uncertain according to science[12] [38].

As Benalcázar and colleagues put it:

> The 2008 constitutional framework offered an unprecedented opportunity to transform national society. The recognition of new forms of coexistence among ethnic groups (interculturality, plurinationality,[13] and collective rights), popular power (participatory democracy and decentralisation), a harmonious relationship between humans and a nature seen as a living transcendental being (the rights of nature), and production formulas (social and solidarity economy, food sovereignty) was undoubtedly a solid starting point [18, p. 165].

Indeed, it was an opening that would lead to a very fertile period for novel legislation produced in a participatory manner. The next section explores this legislative boom, starting with the *Buen Vivir*-inspired development plan and culminating in the 2018 Organic Law of Agrobiodiversity, Seeds, and Promotion of Agriculture.

2009–2018 **Buen Vivir** *legislative boom*

Buen Vivir *development plan (2009–2013)*

This section will list the main objectives of the development plan related to seed system transformation. This document touches upon many dimensions of Ecuadorian society, and it is the basis for the implementation of the citizen's revolution. The main proposition of the plan is a paradigm change from Development to *Buen Vivir*. Many decolonial and post-Development concepts appear in Chapter 3, positioning 'Development' as a product of colonial knowledge and a failed system to produce wellbeing. There are mentions of a Western monoculture making other systems invisible and promoting the idea of an unilinear history. Inequality between North and South nations is emphasised, positioning the consumption levels of rich nations as a certain route towards planetary collapse. The document also produces a critical reading of Ecuador's history and the structural causes of inequality up to the recent decades of neoliberalism [5].

From these critiques of Modern Development and Ecuadorian 'developmentalism', the document goes on to propose a 'displacement of the word development to incorporate the concept of "*Buen Vivir*" into the debate' [5, p. 32]. The concept of *Buen Vivir* can be read as having seven interdependent and overlapping axes in the document: 'intercultural and plurinational state, participatory democracy and social accountability, rights of nature, decentralisation, the social economy, economic sovereignty and the change of the production model, and sovereignty of the body' [18, p. 153]. The document explicitly develops ten *Buen Vivir* principles aligned with the thematic axes, connecting the concept to the rights enshrined in the 2008 constitution.

The development plan can be interpreted as 'anti-developmentalism', pursuing post-neoliberalism through *Buen Vivir*. Among the strategies to achieve this, it proposes: a democratisation of the means of production and new modes of distributing wealth; the transformation of the economy to overcome the extractivist primary-sector export model; and more integration with Latin America.

Crucially, the plan focuses on rural *Buen Vivir*, going beyond an understanding of the rural as agricultural production, and towards an integrated approach to its political economy. The plan recognises rural diversity to develop differentiated strategies for broad inclusion (i.e., recognition of rural women and ethnic differences). From these framings, the plan proposes 12 national objectives for *Buen Vivir*:

i Promote equality, cohesion, and social and territorial integration in diversity.
ii Improve the capabilities and potential of citizens.
iii Improve the population's quality of life.
iv Guarantee the rights of nature and promote a healthy and sustainable environment.
v Guarantee sovereignty and peace and promote strategic insertion in the world and Latin American integration.

vi Guarantee stable, just, and dignified work in its diversity.
vii Build and strengthen public and intercultural spaces of encounter.
viii Affirm and strengthen national identity, diverse identities, plurinationality, and interculturality.
ix Ensure the observance of rights and justice.
x Guarantee access to public and political participation.
xi Establish a social, solidary, and sustainable economic system.
xii Build a democratic state for *Buen Vivir.*

Among the many national territorial strategies, it is important to highlight the strategy to 'promote Buen Vivir and food sovereignty in rural territories' [5, p. 377]. Its main pillar is the renewal of national food production (mainly peasant production) through democratisation of the means of production, promotion of short production chains, diversification, cooperation, and infrastructure. Food sovereignty is mentioned in the plan as a right to healthy, nutritious, and culturally appropriate food, also emphasising that society should decide what, how, and where to produce food. Rural strategies should also include the promotion of agroecology, the recovery of ancestral production knowledge(s) and technologies, and diversification [5].

Organic Law of the Food Sovereignty Regime (2010)

One of the main demands of social movements during the drafting of the constitution was the inclusion of a right to food sovereignty and agrarian reform. However, this was a convoluted process, since multiple actors brought different interpretations of food sovereignty to the table [40].

Food sovereignty was hinted at in the 2009–2013 *Plan Nacional para el Buen Vivir* and was eventually enshrined in the Organic Law of the Food Sovereignty Regime (*Ley Orgánica del Régimen de la Soberanía Alimentaria*, LORSA) with the objective of guaranteeing 'permanent self-sufficiency of healthy, nutritious, and culturally appropriate food to persons, communities, and peoples' [41, p. 3]. This law contemplates: (i) equitable access to water and land, privileging small producers and women heads of households; (ii) the protection of agrobiodiversity; (iii) research, technical assistance, and dialogue of knowledge(s); (iv) participation; and (v) food production and commercialisation.

Regarding the protection of agrobiodiversity, the state commits to 'promot[ing] the recovery, use, conservation, and development of agrobiodiversity and connected ancestral knowledge(s)' [41, p. 4]. Seeds are specifically addressed in Article 8, where 'germplasm, seeds, native plants, and associated ancestral knowledge(s)' [41, p. 5] are established as the patrimony of the Ecuadorian people, and as such, unable to be appropriated through patents or other forms of intellectual property. The law further commits to the protection and promotion of native seed use, conservation, qualification, and free exchange, though not regulating them directly.

Regarding research, technical assistance, and dialogue of knowledge(s), the law aims to improve food nutritional quality, productivity, and safety while enriching agrobiodiversity. It commits to participatory research based on knowledge exchanges with small- and medium-scale producers. Additionally, it advocates for the respect of local biodiversity management practices and the preservation, protection, and development of 'collective knowledge, sciences, technologies, ancestral knowledge, and genetic resources that contain biological diversity and agrobiodiversity' [41, p. 5]. The law commits to the development of institutions, and research and extension programmes for these purposes involving the national science and technology system.

In Articles 14 and 15, the state commits to stimulating agroecological, organic, and sustainable agriculture. This is achieved through diverse financial incentives and state buying programmes giving preference to small and medium agroecological producers. The law declares Ecuador as free from transgenic seeds except in cases of national interest, in which case the president can debate their introduction with the National Assembly. There are also commitments to regulate biotechnological research, observing the precautionary principle when importing commodities of transgenic origin, and labelling transgenic products.

Regarding social participation in food sovereignty, the law states that 'the elaboration of laws and the formulation and implementation of public policies for food sovereignty will have the broadest social participation, through public deliberation processes' [41, p. 8]. For this purpose, it created the Plurinational and Intercultural Conference of Food Sovereignty (COPISA). This body would be formed by state-elected representatives from eight civil society sectors.[14] Its mission would be the development of internal rules, promotion of dialogue, to lead research on food sovereignty, and produce reports to advance legislation.

The law established that COPISA representatives would be elected 90 days after its publication and would have to produce a wide participatory debate process within 180 days of its formation. This process would have to produce laws regulating: (i) the use of and access to land, communes, agrobiodiversity, and seeds; (ii) agricultural development; (iii) agroindustry and agricultural employment; (iv) animal and plant health; and (v) the ability of peasants and Indigenous people to access public credit, insurance, and food subsidies [41, p. 8].

Participatory food sovereignty legal framework (2010–2014)

Between 2010 and 2012, the COPISA organised a participatory process to develop Ecuador's food sovereignty legal framework. COPISA held 188 workshops with more than 15,000 people and 5,000 organisations throughout the country to discuss eight policy proposals concerning: (i) land; (ii) artisanal fishing, aquaculture, and mangrove fishery conservation; (iii) seeds, agrobiodiversity, and agroecology; (iv) ancestral lands and communal property; (v) food safety and quality control; (vi) agricultural development and employment; (vii) credit, subsidies, and insurance; and (viii) consumer health and nutrition [3].

The mechanics of the COPISA process are too extensive to discuss in detail in this book. However, it is important to highlight that it was an instance of participatory and deliberative democracy. Workshops were broadly promoted to ensure participation from diverse food system actors, most were organised in places where historically marginalised actors could participate (e.g., small churches in rural areas), and members of Indigenous and peasant social movements often participated as facilitators [3], [35].

Discussions for the Organic Law of Agrobiodiversity, Seeds, and Promotion of Agriculture were held over 21 workshops with the participation of more than 2,000 people and 500 organisations that included the Ecuadorian seed guardian network [34]. The main policy issues discussed around the Organic Law of Agrobiodiversity, Seeds, and Promotion of Agriculture were: the regulation of conventional seeds, prohibition of GMOs, promotion of agroecology, local seed preservation, agrobiodiversity conservation, and free seed exchanges [3]. This two-year process led to the production of a seed law draft in 2014. For seed guardian Pedro this was, a 'beautiful but imperfect document … that recognised the value of our seed and limited industrial ones, putting controls over them and not ours' [34].

For some authors, COPISA allowed for the institutionalisation of food sovereignty in Ecuador [3], [35]. For others, COPISA never fulfilled its promises of direct democracy, and became a toothless government agency unable to ensure implementation of the food sovereignty legal regime [42]. Be that as it may, by this point the food sovereignty movement in Ecuador had consolidated, leading Indigenous, peasant, and Montubio organisations to form the coalition Red Agraria [3]. This coalition participated in the COPISA workshops and eventually mobilised *iniciativa popular* (the constitutional provision enabling direct democracy) to gain access to the National Assembly and speak in support of the drafted laws [35].

The influence of Red Agraria and the Ecuadorian seed guardian network clearly shows an expansion of recognition and participation in the national food system. However, almost in parallel, the president was mobilising his expanded powers to negotiate a free trade agreement with the European Union, to keep industrialising national agriculture, and eventually 'reshape the participatory [COPISA] into an unleveraged, low-status agency' [42, p. 156].

A new vision for agriculture (2015–2025)

In 2015, the Ecuadorian Ministry of Agriculture (MAG) published guidelines for a new vision of agriculture in Ecuador to 2025 [43]. The vision contemplates a change in the agricultural model towards a sustainable alternative. It prioritises small- and medium-scale producers, arguing for the relevance of cooperative and community work. It focuses on food quality and safety, recognising gender, cultural, social, and productive diversity.

Importantly, it moves towards agroecological technologies and food system actor cooperation.

After the participatory process of COPISA, and as part of this vision for Ecuadorian agriculture, two laws were still pending approval: the Organic Law of Rural Lands and Ancestral Territories, and the Organic Law of Agrobiodiversity, Seeds, and Promotion of Agriculture.

The Organic Law of Rural Lands and Ancestral Territories was approved in 2016 and regulates the possession, property, administration, and redistribution of rural land as a production factor to guarantee food sovereignty. It deals with communal and ancestral land rights and regulation, the conservation of productive soil, and establishes incentives to favour family peasant agriculture and small- and medium-scale food producers. This law is not reviewed in depth here but, suffice it to say, it pursues food sovereignty and attempts to dismantle latifundia through rural land redistribution. Crucially, it continues the trend of focusing on small-scale agriculture and promoting cooperative and collaborative economic land use [44].

The development process of the Organic Law of Agrobiodiversity, Seeds, and Promotion of Agriculture is explored below.

SEED LAW BATTLES (2016–2017)

From the perspective of seed guardians, the 2014 seed law drafted within COPISA opened new spaces for native seeds and was inclusive of their concerns. However, the draft was put on hold until 2016 when a group of assembly members started negotiating a new seed law behind closed doors. An ally of the Ecuadorian seed guardian network in the National Assembly leaked the text of the proposed law. This draft included a declaration that seeds were a strategic resource of the state and imposed mandatory certification of all circulating seeds. These 'negative aspects … filled [seed guardians] with dread … as the state could become the only one to have the final word regarding seeds' [34].

At this point, the Ecuadorian seed guardian network (hereafter referred to as ESGN) was weakened, since the government of Rafael Correa had started 'persecuting NGOs, leading to loss of financing' [34]. Despite the lack of financial resources, the ESGN joined with the National Agroecological Collective (*Colectivo Nacional Agroecológico*) and other civil society organisations to challenge the proposed seed law. Some of these organisations had stopped speaking to each other after the constitutional process and COPISA. However, they came together around the issue of mandatory seed certification.

The ESGN and their allies used the mandatory pre-legislative consultations to mobilise social movement representatives to each territory 'sending [a unified] message to assembly members' [34]. They expressed concerns regarding free circulation of seeds, protection of native, traditional, and peasant seeds, and expanding seed definitions to include associated ancestral knowledge [45]. The law was finally approved in December 2016, responding to the concerns of social movements, primarily that seeds should be declared part of the national cultural heritage and that two parallel seed systems should be created [34].

ORGANIC LAW OF AGROBIODIVERSITY, SEEDS, AND PROMOTION OF AGRICULTURE (2017)

This section provides an overview of the 2017 seed law, focusing on the aspects that deal with informal(ised) seed practices. It views the document as a set of rules that level the playing field for AfroIndopeasant[15] food producers, as the law recognises the role of native seeds and associated ancestral knowledge, sustainable agriculture, and peasant families in the Ecuadorian seed system.

This is evident in the preamble of the law as it aims to:

> protect, revitalise, multiply, and dynamise agrobiodiversity related to plant genetic resources for food and agriculture; ensure the production of, and free and permanent access to, a variety of quality seeds, through promotion and scientific research and the regulation of sustainable agricultural models; respect diverse identities, knowledge(s), and traditions in order to guarantee the self-sufficiency of healthy, diverse, nutritious, and culturally appropriate food to achieve food sovereignty and contribute to Buen Vivir or Sumak Kawsay [46, p. 2].

In addition, the law:

> guarantees the use, production, promotion, conservation, and free exchange of peasant seeds, including native and traditional seeds; and the production, certification, marketing, import, export, and access to certified seeds, through research and promotion of sustainable agriculture [46, p. 2].

Among the principles of the law, it is worth highlighting those relating to sustainability, interculturality, solidarity, and equity. The principle of heritage establishes that native seeds are the cultural heritage of peoples and Indigenous nationalities of Ecuador, which could allow seed guardians to challenge the patenting of native seeds [34].

Among other aspects, the law aims to strengthen the use, preservation, and free exchange of native seeds, clearly distinguishing them from certified seeds. For this purpose, it aims to 'guarantee availability and equity in the access to native, peasant, and certified seed under equal conditions and without discrimination for food security and sovereignty' [46, p. 4]. Although this guideline could be interpreted in many ways, this research frames it as a proposal for a 'level playing field' where both native and industrial seeds have equal opportunities.

The law also establishes several rights that include the 'free production, preservation, commercialisation, exchange, and access to seeds of all kinds' [46, p. 4]. Other rights include the right to participate in decision-making regarding agrobiodiversity, and a right to the recognition and valuing of ancestral and traditional knowledge(s) tied to agrobiodiversity and seed production. The section on rights also recognises the pre-eminent role that women and older adults play in seed conservation and guardianship.[16]

Among the obligations assumed by the state, the main one in relation to informal(ised) seed systems is to

> preserve, produce, regenerate, maintain, revitalise, distribute, promote, and facilitate the use, free exchange and consumption, in a sustainable manner, of native and peasant agrobiodiversity and seeds, the recovery and promotion of knowledge, ancestral knowledges, agroecological and organic practices and technologies linked to them [46, p. 6].

Other state obligations include prohibiting the import of transgenics and appropriation of seed-related knowledge (i.e., patenting collective knowledge of agrobiodiversity), as well as controlling the country's status as free of GMOs. The state would also adopt measures to avoid the loss of traditional knowledge and cultural practices associated with seeds, and guarantee participatory scientific research, development, and innovation within a framework of a 'dialogue of knowledge(s)'.[17] Finally, in what could be associated with the constitution's recognition of nature as an agent, there is an obligation to recognise the intrinsic value of agrobiodiversity [46].

The law recognises two seed production systems: conventional (industrial certified seed) and non-conventional. The non-conventional system is based on peasant seeds, including traditional and native seeds. Native and traditional seeds are clearly defined, and the law calls for the National Agrarian Authority to identify the main native seeds used in Ecuadorian agriculture.

Under the law, native seeds are defined as

> all native or autochthonous sexual and asexual plant reproductive material that maintains its reproductive capacity, that has been domesticated, conserved, raised, cared for, used, and exchanged by producers, communes, communities, peoples, and [Indigenous] nationalities according to their diverse knowledge(s) and cultures, whose use, conservation, classification, exchange, promotion, and protection correspond to individuals and communities with the support of the state [46, p. 20].

One of the law's titles is dedicated to sustainable agriculture, defined as

> agricultural production systems that allow food to be obtained in a stable, healthy, economically viable, and socially acceptable way, in harmony with the environment and preserving the potential of productive natural resources, without compromising the present and future quality of the soil, reducing the risks of environmental degradation, and physical, chemical, and biological contamination of agricultural products [46, p. 14].

The law recognises agroecology, permaculture, and organic and biodynamic agriculture (among other practices) as models of sustainable agriculture. The state commits to preventing plant diseases and epidemics through sustainable practices (without outright banning pesticides), educating the public on the benefits of sustainable agriculture, and promoting peasant, family, and community agriculture to energise this agricultural model; as the law frames it, to 'recover the talent, prominence, and centrality of the peasant family in the development of agrarian and food systems' [46, p. 15].

The 2017 seed law recognises the 'non-conventional' seed system. It includes and promotes native seeds and protects associated traditional knowledge. However, the law was not without controversy; the ESGN and their allies brought a suit against 15 of its articles, which would trigger a rewrite if approved. Additionally, two months before leaving the presidency, Rafael Correa vetoed Article 56 allowing transgenics into the country for research purposes [34]. Despite the difficulties of reconciling industrial agriculture with alternative practices, and reconciling certified and peasant seeds, the law is meticulous in its definitions and provisions. It connects to the lineage of the *Buen Vivir*-inspired constitution with its considerations of nature, expanding participation, and food sovereignty to provide a template for a future seed system.

Seed system after the legislative boom

Despite formulating a level playing field for all system actors in the 2017 seed law, seed battles raged on. Tensions between the 'conventional' and 'non-conventional' seed system continue, but novel interactions provide clues of a Transmodern system emerging. This section highlights some activities of the Ecuadorian seed guardian network (ESGN) and the Ecuadorian Ministry of Agriculture and Livestock after the *Buen Vivir* legislative boom.

The Ecuadorian seed guardian network

The ESGN ramped up its activities after the signing of the Organic Law of Agrobiodiversity, Seeds, and Promotion of Agriculture. In addition to the legal challenge to 15 of the law's articles, the ESGN wanted to contest government assertions that organic seed production was non-existent in Ecuador. As part of their advocacy, the ESGN created an association of organic seed producers and registered a company to sell their seeds. This made visible the 'non-conventional' seed system to the Superintendency of Control over Market Power (*Superintendencia de Control de Poder en Mercado*, SCPM), allowing seed guardians to be heard in court and gain access to proposed seed regulations.

The SCPM ruled that the 2017 seed law produced market disparity, since it stated that there are two seed markets (one commercial and the other non-commercial). This opened ground for an unfair competition case, since the two seed systems are part of the same market, and state support for commercialisation is only given to the 'conventional' system. The case did not advance due to a change of government, but it was an important moment for the ESGN [34].

A change in strategy for the ESGN was brought in 2018, favouring more outreach activities through social media. The network realised that they had 'more than 3,000 seed varieties preserved, but they were going to die with [them] if nobody else learned [about them]' [34]. The network hired a social media manager, started the Radio Semilla (Radio Seed) podcast, and published actively on Facebook to reach a wider audience. The ESGN assembly also created a web-based platform called *Madre Semilla* (Mother Seed) on which it offers massive open online courses on seed care, agroecology, permaculture, cooking, and associated topics [47]. The platform also holds printed and virtual resources, a documentary series about seed guardianship and alternative food practices, and a virtual store. Most of these are paid offerings, and revenues are directed towards network activities.

The ESGN has maintained its annual meetings since the signing of the seed law in 2017. A particular highlight of ESGN events was the 2019 virtual 'seed festival' that had to move online due to the COVID-19 pandemic. The network was planning an event for 2,000 people but ended up attracting more than 18,000 from all over Latin America. The ESGN realises that their audience numbers are minimal compared to traditional media, but they 'feel they are doing the work encouraging people to get on a transformation pathway' [34].

The network believes that a 'state that listens' and 'social dialogue' are essential for transformations to happen. However, they feel that old power asymmetries are still present in the Ecuadorian seed system. In one guardian's words: 'We had to go to trial with the Ministry of Agriculture so that they would show us information regarding the regulations they were developing. Meanwhile, the big capitalist corporation can call the minister directly each week' [34].

Looking towards the future, the ESGN will maintain their efforts to be an interlocutor of the state to produce change. However, they are acutely aware of the difficulties of achieving a truly open dialogue; in the meantime, they will keep trying to create and implement alternative systems and diffuse them [34].

State interventions in the non-conventional system

This section focuses on the activities of the Ecuadorian Ministry of Agriculture and Livestock (*Ministerio de Agricultura y Ganadería*, MAG)[18] during and after the drafting of the 2017 seed law. Its engagement with the national seed system has been constant before, during, and after the citizen's revolution (e.g., the 2011 plan for the promotion of quality seed). The intention of this section is to highlight some activities that evidence increased support for the non-conventional seed system.

It is important to recognise that there is still a heavy agro-industrial bias in the MAG, as has been mentioned in previous sections. For instance, in 2017 the MAG implemented the Great Agrarian *Minga*,[19] which aimed to provide farmers with property titles for their land, loans, irrigation, seed kits, insurance, technical assistance, agricultural mechanisation, and access to markets. This programme was widely criticised, since it benefited mostly big landowners, delivered certified hybrid seeds with pesticides in its seed kits, and distributed them through big commercial enterprises [48]. As recently as 2022, the MAG implemented a 'seed plan' benefiting growers of multiple food products in a project that invested more than US$14 million to deliver certified seeds and associated chemical fertilisers and pesticides [49].

The MAG and seed exchanges

The MAG has participated in local seed fairs since 2013, when the vice-minister of rural development attended the Muyu Rami fair organised by the Union of Peasant and Indigenous Organisations of Cotacachi (*Unión de Organizaciones Campesinas Indígenas de Cotacachi*, UNORCAC). However, MAG participation in seed and knowledge exchanges appears to increase from 2016, as it also assumed a more active role in organising them.

In 2016, regional MAG representatives participated in the Second Seed Exchange and Agrobiodiversity Fair in Catacocha, Loja. In 2017, the ministry co-organised the Native Seed Exposition and Peasant and Family Agriculture Fair in Sigsig. This event was touted as an effort to establish short commercialisation chains while also offering locally produced organic fertilisers. In 2018, the MAG co-organised the Second National Seed Exchange Killa Raymi celebrated during the ancestral Andean festival Killa Raymi, where tribute to fertility and seeds is given. All seeds exchanged were agroecological, and the event was framed as an act to rescue ancestral knowledge (s) and millenary communitarian practices [50].

In 2019, the MAG organised the First Seed and Ancestral Knowledge Exchange Kapak Raymi coinciding with a pre-Hispanic celebration in honour of the sun. The event was framed as an intercultural and intergenerational encounter where organic seeds were exchanged and included an ancestral ritual to thank Pachamama. These events have been constant up through 2025, with the Seventh Seed and Ancestral Knowledge Exchange Fair – Kulla Raymi [51]. Increased attention to Indigenous practices in MAG-organised seed fairs might correlate with the creation of a directorate of ancestral knowledge in the ministry (discussed further below).

Peasant and Family Agriculture (PFA)

In 2017, the MAG created the division of Undersecretary of Peasant and Family Agriculture through a presidential directive. This unit supports the transition from conventional agriculture to more holistic practices, and strengthens peasant and family agriculture through loans, subsidies, and other strategies. The PFA division has four directorates in its structure, importantly a Directorate of Agroecological and Sustainable Production, and a Directorate of Ancestral Knowledge. The latter's objectives include:

- Promoting the use and consumption by producing families of different seed varieties adapted to each particular area.
- Promoting exchanges of native seeds between producing families.
- Increasing local native seeds to benefit other producing families.
- Increasing arable areas and local yields of native and traditional seeds.
- Promoting the creation of local native seed houses and interconnections between them.
- Encouraging the creation of a network of seed growers at local, provincial, and national levels [52].

In late 2017, the MAG launched a PFA seal as a strategy to communicate with consumers about the social origin of food products from peasant and family production units. Individuals and organisations can apply for the PFA seal at regional MAG offices at no charge, and after a technical visit they might get a certificate and a style guide on how to use the seal on their products. Despite its commitment to PFA and actions like the PFA seal, the Undersecretary of Peasant and Family Agriculture has not been as effective as social movements would have hoped.

This might be due to the small number of employees compared to the overall size of the MAG and budget cuts that have undermined mission fulfilment [53]. It is important to highlight that the Food and Agriculture Organisation of the United Nations (FAO-UN) declared 2019 to 2028 as the UN Decade of Family Farming. This initiative aims to situate family farming at the centre of public policy and investment, so there might be more support to the Ecuadorian MAG PFA unit in following years.

In 2020, the First International Summit of Peasant Family Agriculture was organised by the National Confederation of Peasant, Indigenous, and Black Organisations (*Confederación Nacional de Organizaciones Campesina, Indígenas y Negras*, FENOCIN), and the Union of Intercultural Social Organisations of Pichincha (*Unión de Organizaciones Sociales Interculturales de Pichincha*, UOSISP). Both FAO-UN and the MAG made presentations at this summit, alongside social organisations and state representatives from several Latin American countries.

At the event, the Ecuadorian Minister of Agriculture and Livestock presented the PFA strategies and framed PFA challenges as promoting quality and traceability in PFA chains, promoting rural innovation and digitisation, opening markets for PFA products, and closing the gap with consumers through awareness campaigns for healthy and responsible consumption [54]. As part of the summit, the MAG declared 30 October as the National Day of

Peasant Family Agriculture, an annual celebration to recognise the importance of PFA in the Ecuadorian food system.

Increased visibility of native seeds in the MAG

In addition to MAG-sponsored seed meetings and the increased representation of PFA and ancestral knowledge in the organisation, there is also evidence of knowledge exchanges between native seed guardians and MAG. In 2020, ministry representatives made an 'observation tour' around the country, visiting producers of peasant seeds [52], e.g., visiting the UNORCAC Bio-knowledge Centre to understand processes of native seed conservation and develop technical assistance strategies. The MAG also discussed with them the possibility of establishing a seed bank at the Bio-knowledge Centre dedicated solely to native seeds [55].

In 2021, the Undersecretary of Peasant and Family Agriculture taught a Training Course for Rural Extension Agents in Sustainable Agriculture to MAG technicians. A module on Production, Extraction and Conservation of Native and Traditional Seeds was part of the course, reaching all district directorates [56]. This course aimed to strengthen technicians' knowledge of native seeds, and to show the process of extraction and conservation of peasant seeds, and how to implement native seed houses [52]. The presentation given during this module shows a wide range of native seed houses, Andean rituals as part of seed exchanges, and establishes step-by-step techniques and procedures for peasant seed extraction, conservation, and seed house implementation.

The MAG continues to support spaces of dialogue of knowledge(s) connecting PFA actors and technicians to share native seeds and associated methods of care.

Conclusion

The recent transformation of the Ecuadorian seed system can be understood as a nested process 30 years in the making. It began in the early 1990s with the emergence of the Quichua neologism *Sumak Kawsay* as both an alternative vision of the future and a platform to achieve it. This Transmodern narrative opened a space to cross the abyssal line, allowing marginalised actors to revalue their ways of life and position their life-worlds as futural.[20] Late 20th century strengthening of rural social movements, especially Indigenous, moved *Buen Vivir/Sumak Kawsay* from the periphery to the centre of Ecuadorian politics.

After decades of political turmoil, a '21st century socialist' project emerged as an 'outsider' response to the environmental, social, and political crisis. However, it was only through its integration with social movements and the narrative of *Buen Vivir* as an alternative to development that it came to power in the early 2000s.

Once in power, this '*movimientista* left' (social movement-aligned left) set out to implement a citizen's revolution that overhauled citizen participation and directed political institutions towards *Buen Vivir* instead of developmentalism. The next decade (2008–2018) was a fertile time for novel legislation to emerge, beginning with a constitution that recognised the rights of nature and rights of people to food sovereignty, and which led to the inevitable transformation of food system rules. The post-constitutional Organic Law of Food Sovereignty opened a broad participatory process (COPISA) to include historically marginalised voices in the transformation of food systems.

The Ecuadorian seed guardian network (ESGN) has participated in each step of this process, from the 2008 constitutional assembly up to still-unresolved challenges to the Organic Law of Agrobiodiversity, Seeds, and Promotion of Agriculture. New system rules opened more spaces for non-Modern and alternative practices of food production; they recognised AfroIndopeasant knowledge, and the importance of peasant family agriculture in the food system. Despite the persistent agro-industrial bias in Ecuadorian state agencies, and all the failed implementations of a *Buen Vivir* vision, this legislative process levelled the playing field somewhat to give more space to alternatives.

Through their activities over the last 20 years, the ESGN and their allies became influential players in the national food system, allowing for non-Modern narratives and practices to be seen. The Ecuadorian legislative process is Transmodern, as it opened several 'translational contact zones' where the Modern and non-Modern engaged in dialogue, agonism, and antagonism to hybridise.

The resulting hybrids (the laws themselves and resulting strategies) are Transmodern objects, as they include nature, recognise interrelatedness, and create a scaffolding for non-Modern, earth-centred, and post-capitalist alternatives to grow. The 'sustainable materialism'[21] of the ESGN continues, as does its advocacy pressuring institutions to 'see' the non-conventional seed system. Meanwhile, the Undersecretary of Peasant and Family Agriculture and its Directorate of Ancestral Knowledge work inside the state apparatus to promote more horizontal 'dialogues of knowledge(s)' in favour of *Buen*(*os*) *Vivir*(*es*).

Notes

1 Here, the term 'line of flight' ('ligne de fuite' in the original French) comes from the work of Deleuze and Guattari.
2 The Andean Community of Nations comprises Bolivia, Colombia, Ecuador, Peru, and Venezuela.
3 For a notable Indigenous political movement review the Zapatista revolution in Mexico.
4 This recent pursuit of alternatives to capitalism might be inspired by José Carlos Mariátegui's proposal for an Indo-American socialism.
5 These values might be inspired by Catholic and Protestant liberation theology.

6 For example, through the Bolivarian Alliance for the Peoples of Our America (*Alianza Bolivariana para los Pueblos de Nuestra América*, ALBA).
7 For example, the constant references to the independence-era figure Simón Bolívar in post-1999 Venezuela, after whom Venezuela's Bolivarian Revolution – and the Bolivarian Alliance – are named.
8 For an example, see the proposals developed by CONAIE [57].
9 Pachamama is the Inca/Andean mother-like goddess who creates and sustains life on earth, aids in planting and harvesting, and represents fertility.
10 From the Spanish *calidez*, which in this context refers to human warmth.
11 Montubio peoples are mestizos who live on the Ecuadorian coast and are recognised as a distinct ethnicity.
12 This is a veiled form of speaking about genetically modified organisms.
13 This term is used to recognise that several 'national communities' with their own culture coexist within the state.
14 The eight sectors represented are: universities/researchers; consumers; small/medium-sized industrial producers; small/medium-sized farmers; small/medium-sized cattle ranchers; artisanal fishers and aquaculture; peasants and irrigators; and Indigenous, Afro-Ecuadorian, and Montubio communities.
15 Under the 2017 seed law, the term AfroIndopeasant includes Montubiuo actors.
16 This is reminiscent of the concept of *maestres pueblo* used in Venezuelan seed legislation.
17 This 'dialogue of knowledge(s)' calls for a respect for ancestral wisdom related to agrobiodiversity, seeds, and sustainable agriculture. It recognises a plurality of knowledge and prioritises young and female technicians from the areas where the participatory innovation process happens.
18 MAG was formerly the Ministry of Agriculture, Livestock, Aquaculture, and Fishing (Ministerio de Agricultura, Ganadería, Acuacultura y Pesca, MAGAP).
19 'Minga' is an Indigenous Andean term that refers to collaborative engagements between family, friends, and neighbours to benefit the community.
20 The term 'futural' refers to producing futures, in opposition to the defuturing produced by modern society that takes futures away from humanity. For more on these concepts see [58].
21 The term 'sustainable materialism' refers to the networked sustainable practices of everyday life as a form of activism (in this case food practices). See [59] for more.

References

[1] M. Tanasescu, 'The rights of nature in Ecuador: The making of an idea', *International Journal of Environmental Studies*, vol. 70, no. 6, pp. 846–861, Dec.2013, doi:10.1080/00207233.2013.845715.

[2] P. Clark, 'Can the state foster food sovereignty? Insights from the case of Ecuador', *Journal of Agrarian Change*, vol. 16, no. 2, pp. 183–205, Apr.2016, doi:10.1111/joac.12094.

[3] K. Peña, 'Institutionalizing food sovereignty in Ecuador', in *Food Sovereignty: A Critical Dialogue*, 2013. [Online]. Available: https://www.iss.nl/en/media/51pena2013-1.

[4] P. Clark, 'Neo-developmentalism and a "vía campesina" for rural development: Unreconciled projects in Ecuador's Citizen's Revolution', *Journal of Agrarian Change*, vol. 17, no. 2, pp. 348–364, Apr.2017, doi:10.1111/joac.12203.

[5] Ecuador. Secretaría Nacional de Planificación y Desarrollo. (2009). *Plan nacional para el Buen Vivir, 2009–2013*. Accessed: 10 Feb. 2026. [Online]. Available: https://www.ecuadorencifras.gob.ec/wp-content/descargas/Informacion-Legal/Normas-de-Regulacion/Plan-Nacional-para-el-Buen-Vivir/Plan+Nacional+del+Buen+Vivir+2009-2013.pdf.

[6] A. Acosta, 'El *Buen Vivir* como alternativa al desarrollo Algunas reflexiones económicas y no tan económicas', *Politica y Sociedad*, vol. 52, no. 2, pp. 299–330, 2015, doi:10.5209/rev-POSO.2015.v52.n2.45203.
[7] A. L. Hidalgo-Capitán and A. P. Cubillo-Guevara, 'Deconstruction and genealogy of Latin American good living (*Buen Vivir*). The (Triune) good living and its diverse intellectual wellsprings', in *Alternative Pathways to Sustainable Development: Lessons from Latin America*, G. Carbonnier, H. Campodónico, and S. T. Campodónico, Eds. Leiden: Brill Nijhoff, 2017, doi:10.1163/9789004351677.
[8] F. Huanacuni Mamani, *Buen Vivir/Vivir Bien Filosofía, políticas, estrategias y experiencias regionales andinas.* Lima: Coordinadora Andina de Organizaciones Indígenas (CAOI), 2010. [Online]. Available: https://www.reflectiongroup.org/stuff/Vivir-bien.
[9] S. Schavelzon, 'El concepto de Vivir Bien/Buen Vivir', in *Plurinacionalidad y Vivir Bien/Buen Vivir.* Buenos Aires: CLACSO, 2015, pp. 181–267.
[10] C. Walsh, *Interculturalidad, estado, sociedad. Luchas (de)coloniales de nuestra época*. Quito: Ediciones Abya-Yala, 2009.
[11] A. Acosta, '*Buen Vivir*: A perspective for rethinking the world', in *Degrowth in movement(s): Exploring pathways for transformation*, C. Burkhart, N. Treu, and M. Schmelzer, Eds. Winchester: Zero Books, 2021, Ch. 5.
[12] F. García *et al.*, 'El *Buen Vivir* como paradigma societal alternativo', *Economistas sin Fronteras*, 2016.
[13] P. Altmann, 'Studying discourse innovations: The case of the indigenous movement in Ecuador', *Historical Social Research*, vol. 40, no. 3, pp. 161–184, 2015, doi:10.12759/hsr.40.2015.3.161-184.
[14] A. E. Beling, J. Vanhulst, F. Demaria, V. Rabi, A. E. Carballo, and J. Pelenc, 'Discursive synergies for a "great transformation" towards sustainability: Pragmatic contributions to a necessary dialogue between human development, degrowth, and *Buen Vivir*', *Ecological Economics*, vol. 144, March, pp. 304–313, 2018, doi:10.1016/j.ecolecon.2017.08.025.
[15] N. Chassagne, 'Sustaining the "good life": *Buen Vivir* as an alternative to sustainable development', *Community Dev J*, vol. 54, no. 3, pp. 482–500, 2019, doi:10.1093/cdj/bsx062.
[16] E. Gudynas, '*Buen Vivir*: Today's tomorrow', *Development*, vol. 54, no. 4, pp. 441–447, 2011, doi:10.1057/dev.2011.86.
[17] P. Solón, 'Vivir Bien: Old cosmovisions and new paradigms', February2018. Available: https://greattransition.org/publication/vivir-bien.
[18] P. C. Benalcázar and F. J. U. de la Rosa, 'The *Buen Vivir* postdevelopmentalist paradigm under Ecuador's citizens' revolution governments (2007–2017): An appraisal', *Lat Am Perspect*, vol. 48, no. 3, pp. 152–171, May2021, doi:10.1177/0094582X211004910.
[19] J. Polo Blanco and E. Piñeiro Aguiar, 'El *Buen Vivir* como discurso contrahegemónico. Postdesarrollo, indigenismo y naturaleza desde la visión andina', *Mana*, vol. 26, no. 1, pp. 1–31, 2020, doi:10.1590/1678-49442020v26n1a205.
[20] A. Acosta, 'Post-growth and post-extractivism: Two sides of the same cultural transformation', *Alternautas*, vol. 3, no. 1, pp. 50–71, 2016.
[21] G. G. Prudencio and J. P. Blanco, 'La construcción discursiva del *Sumak Kawsay* Ecuatoriano y su relación con la consecución de los objetivos del *Buen Vivir*', *Ánalisis Político*, vol. 30, no. 89, pp. 76–90, 2017.
[22] A. V. Recasens, '"Pachamamista" discourses versus development policies: The debate over *Sumak Kawsay* in the Andes', *Revista de Ciencias Sociales*, vol. 48, pp. 55–72, 2014.

[23] E. Ranta, ‘Toward human/conviviality: *Buen Vivir* as a transformative alternative to capitalist Coloniality’, in *The Routledge handbook of transformative global studies*, 1st ed., S. A. Hamed Hosseini, J. Goodman, S. C. Motta, and B. K. Gills, Eds. London: Routledge, 2015, pp. 419–430.
[24] W. Mignolo and C. Walsh, *On decoloniality*. Durham, NC: Duke University Press, 2020. doi:10.1515/9780822371779.
[25] W. Sachs, *The development dictionary: A guide to knowledge as power*, 2nd ed., vol. 18, no. 1. London: Zed Books, 2010. doi:10.2307/1971880.
[26] A. Ziai, ‘Post-development 25 years after *The Development Dictionary*’, *Third World Q*, vol. 38, no. 12, pp. 2547–2558, 2017, doi:10.1080/01436597.2017.1383853.
[27] K. Weld, ‘Holy War: Latin America’s far right’, *Dissent*, 2020. [Online]. Available: https://www.dissentmagazine.org/issue/spring-2020.
[28] A. P. Cubillo-Guevara and A. L. Hidalgo-Capitán, ‘Good living goals. A proposal for the construction of a global transdevelopment’, *Alternautas*, vol. 6, no. 2, pp. 71–87, 2019.
[29] P. R. Andrade, ‘El reino (de lo) imaginario: Los intelectuales políticos ecuatorianos en la construcción de la Constitución de 2008’, *Ecuad Debate*, vol. 85, pp. 35–48, 2012. [Online]. Available: https://www.researchgate.net/publication/278005957.
[30] S. Ellner, ‘The distinguishing features of Latin America’s new left in power: The Chávez, Morales, and Correa governments’, *Lat Am Perspect*, vol. 39, no. 1, pp. 96–114, Jan.2012, doi:10.1177/0094582X11425333.
[31] H. Salgado Pesantes, ‘El proceso constituyente del Ecuador. Algunas reflexiones’, *Revista IIDH*, vol. 47, pp. 205–223, 2008.
[32] J. J. Paz and M. Cepeda, ‘La Asamblea Constituyente de 2007: Un nuevo ciclo histórico en Ecuador’, constitutionnet.org. Accessed: 15 Dec. 2022. [Online]. Available: https://constitutionnet.org/sites/default/files/ciclo_historico.pdf.
[33] H. Wittman, ‘From protest to policy: The challenges of institutionalizing food sovereignty’, *Canadian Food Studies/La Revue canadienne des études sur l’alimentation*, vol. 2, no. 2, pp. 174–182, Sep.2015, doi:10.15353/cfs-rcea.v2i2.99.
[34] NET006, ‘private communication’, Nov.2021.
[35] K. Peña, ‘Social movements, the state, and the making of food sovereignty in Ecuador’, *Lat Am Perspect*, vol. 43, no. 1, pp. 221–237, Jan.2016, doi:10.1177/0094582X15571278.
[36] M. Becker, ‘Correa, Indigenous movements, and the writing of a new constitution in Ecuador’, *Lat Am Perspect*, vol. 38, no. 1, pp. 47–62, Jan.2011, doi:10.1177/0094582X10384209.
[37] E. Gudynas, ‘La ecología política del giro biocéntrico en la nueva Constitución de Ecuador’, *Revista de Estudios Sociales*, vol. 32, pp. 34–46, 2009. [Online]. Available: http://www.redalyc.org/articulo.oa?id=81511766003.
[38] Ecuador. Asamblea Constituyente. (2008). *Constitución del Ecuador*. Accessed: 10 Feb. 2026. [Online]. Available: www.oas.org/juridico/pdfs/mesicic4_ecu_const.pdf.
[39] M. de la Cadena, ‘Indigenous cosmopolitics in the andes: Conceptual reflections beyond “politics”’, *Cultural Anthropology*, vol. 25, no. 2, pp. 334–370, 2010, doi:10.1111/j.1548-1360.2010.01061.x.
[40] B. McKay, R. Nehring, and M. Walsh-Dilley, ‘The “state” of food sovereignty in Latin America: Political projects and alternative pathways in Venezuela, Ecuador and Bolivia’, *Journal of Peasant Studies*, vol. 41, no. 6, pp. 1175–1200, Nov. 2014, doi:10.1080/03066150.2014.964217.

[41] Ecuador. Asamblea Nacional de la República de Ecuador. (2009). *Ley Orgánica del Régimen de la Soberanía Alimentaria.* Accessed: 10 Feb. 2026. [Online]. Available: https://faolex.fao.org/docs/pdf/ecu88076.pdf.

[42] E. Fiorini, *COPISA in Ecuador: Participation that wasn't.* Tucson, AZ: University of Arizona, 2015. [Online]. Available: https://repository.arizona.edu/bitstream/handle/10150/595609/azu_etd_14235_sip1_m.pdf?sequence=1&isAllowed=y.

[43] Ecuador. Ministerio de Agricultura, Ganadería y Pesca. (2016). 'La Política Agropecuaria Ecuatoriana – Hacia el Desarrollo Territorial Rural Sostenible 2015–2025', Ministerio de Agricultura, Ganadería y Pesca [Ecuador]. Accessed: 10 Feb. 2026. [Online]. Available: https://faolex.fao.org/docs/pdf/ecu183434.pdf.

[44] Ecuador. Asamblea Nacional de la República de Ecuador. (2016). *Ley Orgánica de Tierras Rurales y Territorios Ancestrales.* [Online]. Available: www.habitatyvivienda.gob.ec/wp-content/uploads/downloads/2017/11/Ley-Organica-de-Tierras-Rurales-y-Territorios-Ancestrales.pdf.

[45] Ecuador. Asamblea Nacional de la República del Ecuador. (2016).'Soberanía Alimentaria se prepara para las audiencias públicas de la Ley de Semillas'. Accessed: 26 Dec. 2022. [Online]. Available: https://www.asambleanacional.gob.ec/es/noticia/46312-soberania-alimentaria-se-prepara-para-las-audiencias.

[46] Ecuador. Asamblea Nacional de la República de Ecuador. (2017). *Ley Orgánica de Agrobiodiversidad, Semillas y Fomento de Agricultura,*. [Online]. Available: https://www.gob.ec/sites/default/files/regulations/2018-11/Ley-Orgánica-de-Agrobiodiversidad-Semillas-y-Fomento-de-la-Agricultura-Sustentable.pdf.

[47] Madre Semilla, 'Madre Semilla Plataforma Educativa'. Madresemilla. Accessed: 10 Feb. 2026. [Online.] Available: www.madresemilla.com.

[48] X. León, 'La minga agropecuaria', *Ecuador Today*, n.d. Accessed: 10 Feb. 2026. [Online]. Available: https://ecuadortoday.media/2020/10/10/la-minga-agropecuaria.

[49] Ecuador. Ministerio de Agricultura y Ganadería. (n.d.). 'MAG inicia el proyecto semillas que favorece a más de 20 mil productores', Agricultura.gob.ec. Accessed: 10 Feb. 2026. [Online]. Available: www.agricultura.gob.ec/mag-inicia-el-proyecto-semillas-que-favorece-a-mas-de-20-mil-productores/ https://www.agricultura.gob.ec/mag-inicia-el-proyecto-semillas-que-favorece-a-mas-de-20-mil-productores/

[50] Ecuador. Ministerio de Agricultura y Ganadería. (n.d.). 'Intercambio de semillas y fiesta de fertilidad de la tierra se realizará en Cuenca', Agricultura.gob.ec. Accessed: 10 Feb. 2026. [Online]. Available: www.agricultura.gob.ec/intercambio-de-semillas-y-fiesta-de-fertilidad-de-la-tierra-se-realizara-en-cuenca/.

[51] Ecuador. Ministerio de Agricultura, Ganadería y Pesca. (2025). 'Más de 800 personas de Tungurahua participaron del "Kulla Raymi y Séptima Feria de Intercambio de Semillas y Saberes Ancestrales"'. Accessed: 11 Dec. 2025. [Online]. Available: https://www.agricultura.gob.ec/mas-de-800-personas-de-tungurahua-participaron-del-kulla-raymi-y-septima-feria-de-intercambio-de-semillas-y-saberes-ancestrales/.

[52] Ecuador. Ministerio de Agricultura, Ganadería y Pesca. (2021). 'Producción, extracción y conservación de semillas nativas y tradicionales'. Accessed: 10 Feb. 2026. [Online]. Available: www.scribd.com/document/495219290/Produccion-Extraccion-y-Conservacion-de-Semillas-Nativas.

[53] V. Recalde Vela, 'Agricultura familiar campesina comunitaria en Ecuador – Situación, desafíos y oportunidades', 2021. [Online]. Available: https://ocaru.org.ec/en/agricultura-familiar-campesina-comunitaria-en-ecuador-situacion-desafios-y-oportunidades/.

[54] Ecuador. Ministerio de Agricultura, Ganadería y Pesca. (n.d.). 'En Cumbre Internacional, MAG presenta las acciones para impulsar la agricultura familiar

campesina'. Accessed: 10 Feb. 2026. [Online]. Available: www.agricultura.gob.ec/en-cumbre-internacional-mag-presenta-las-acciones-para-impulsar-la-agricultura-familiar-campesina.

[55] Ecuador. Ministerio de Agricultura, Ganadería y Pesca. (n.d.). 'MAG trabaja en el fortalecimiento de sistema de producción de semillas nativas y tradicionales'. Accessed: 10 Feb. 2026. [Online]. Available: www.agricultura.gob.ec/mag-trabaja-en-el-fortalecimiento-de-sistema-de-produccion-de-semillas-nativas-y-tradicionales.

[56] Ecuador. Ministerio de Agricultura, Ganadería y Pesca. (n.d.). 'Técnicos del MAG refuerzan conocimientos para conservar semillas nativas y tradicionales'. Accessed: 10 Feb. 2026. [Online]. Available: www.agricultura.gob.ec/tecnicos-del-mag-refuerzan-conocimientos-para-conservar-semillas-nativas-y-tradicionales.

[57] CONAIE, 'Propuesta de la CONAIE frente a la Asamblea Constituyente', 2007. Accessed: 10 Feb. 2026. [Online]. Available: https://www.yachana.org/earchivo/conaie/ConaiePropuestaAsamblea.pdf.

[58] T. Fry, *Design futuring*. Sydney: University of New South Wales Press, 2009.

[59] D. Schlosberg, 'From postmaterialism to sustainable materialism: The environmental politics of practice-based movements', *Env Polit*, vol. 34, no. 1, pp. 92–112, 2025, doi:10.1080/09644016.2019.1587215.

Part 3

How change happened

7 Translational contact zones emerging in Andean seed systems

How did seed system transformations occur across the abyssal line in Colombia, Ecuador, and Venezuela? Early 21st century transformations evidence the emergence of new Transmodern configurations that allow the Modern and non-Modern to co-create new ways of doing things. In these processes, novel narratives and practices stimulated society–state interactions that produced new sets of rules, effectively Transmodernising national seed systems.

Chapters 4–6 presented chronological narratives of these change processes and should be considered as the empirical level of the object of analysis. Chapter 1 argued that Andean seed systems are fractured by the abyssal line. Divisions between a formal and informal(ised) system are the product of historical monopolisation of knowledge that rendered some knowledge(s) incommensurable and incomprehensible to the Modern/Colonial core. Seed systems show how Modern institutions like capitalism, law, and scientific knowledge now have a monopoly on truth. Accordingly, non-productivist practices and seed bartering are seen as anomalies, non-Modern seed production was kept in a non-legal space, and traditional knowledge/technology was rendered invisible. This chapter focuses on the epistemic, legal, and scientific abyssal lines [1] crossing seed systems and the ways in which actors transcended them.

Andean seed system actors mobilised their epistemological and democratic imagination to create new constellations of knowledge(s) and practices that cross these abyssal lines to build viable alternatives. These time-spaces of negotiation where relational inequality is reduced have been framed in this book as translational contact zones (TCZ). Translation dynamics occurred between both sides of the abyssal line, leading to the recognition of reciprocal incompleteness, allowing the expansion of parameters that define authorship, intelligibility, and purposefulness. This is a fundamental part of an epistemology of seeing, a stance that embraces a plurality of knowledge(s) and practices aiming towards interknowledge – an integrative hybridising process that creates new meaning articulations through bricolage [1].

Seed guardian networks (SGNs) were essential for the opening of these TCZ. SGNs were presented in Chapter 3 as peripheral Transmodern spaces

DOI: 10.4324/9781003715092-11

where non-Modern food practices are strengthened and reproduced. Over a period of 20 years, these networks became influential players in national seed systems, participating in legislative change to enable a more level playing field with formal agricultural practices. This is no small feat, since current food systems perpetuate Coloniality by framing AfroIndopeasant practices as antiquated; an obstacle to progress and becoming fully Modern.

It is important to reiterate that SGNs do not advocate for a return to an idealised past or maintaining traditional food production in a static state. They are exploring new avenues to strengthen these practices, frequently interacting with new technologies and contemporary food science. SGNs are knowledge and technology producers, innovating in a Transmodern fashion as they hybridise knowledge(s).

One of the intentions of this book was to better understand how large system change occurs. After analysing the data while reflecting on this question, three dimensions emerged as the pre-eminent sites where change happened: narratives, practices, and rules. In this chapter, multiple theories will be deployed as lenses to examine early 21st century Andean seed system change. New institutionalism, theories of narrative transformation, and social practice theory (SPT) provide insight on the commonalities between change processes. This book does not attempt to integrate these theories, as it recognises deep differences between them. However, it notes overlaps that allow for a more robust understanding of how change happened.

This chapter aims to provide a description of how translational contact zones opened in Andean seed systems. It does not attempt to prescribe causal relations between phenomena but rather attest to the complexity of interrelations that influenced transformation. The chapter is divided into three sections: (i) Transmodern discourse coalitions present the ideational elements that provided meaning to the change process; they analyse how actors came together through three overlapping narratives – *Buen Vivir*, food sovereignty, and the new plural left; (ii) converging practices argues that the scaling up of SGN practices and state moves towards participatory democracy were essential for systems change; and finally, (iii) changing rules of the game frames the blurring of the legal abyssal line and the emergence of new institutions as a reflection of shifting narratives and practices. These dimensions are not isolated; they should be seen as always interrelated, influenced by each other, and in flux.

Transmodern discourse coalitions

Transformation initiatives intentionally co-create narratives and future visions to assert their purpose and inform action [2]. Narratives, in turn, attract and articulate actors that further reproduce these transformative visions [3]. Discourses and discourse coalitions were identified as essential forces in Adean seed system change. A multiplicity of actors mobilised narratives at national and transnational scales, forming broad coalitions due to shared ideational elements.

Discourses give meaning to physical and social realities, define common sense, legitimate knowledge, and provide the basic terms for analysis, debates, agreements, and disagreements. They are important because they inform problem description, define what is imaginable and possible, and condition the way in which actors perceive the rationale and scope of action [4], [5]. Discourse coalitions are ensembles of actors that are attracted to specific storylines and reproduce them through practices [3].

In Andean seed systems, discourse coalitions were formed through the linking of shared ideational elements between the narratives of food sovereignty, *Buen Vivir*, and the new plural left. These coalitions were instrumental in the transformation of seed systems, as they allowed actors to connect across the abyssal line at different scales. Ultimately, commonality between ideational elements favoured the opening of translational contact zones (TCZ) where agents, knowledges, and practices from both sides interacted [1].

There is an increased interest in understanding the role of ideational elements in processes of systemic change [3], [6], [7]. **Discourses** are defined as a 'web of meanings, ideas, interactions and practices ... expressed or represented in texts within institutional and everyday settings' [3, p. 101]. A discourse is a shared way of apprehending the world [4] that affords and constrains action and behaviour [5]. As such, it crystalises power by promoting or suppressing specific values, identities, relations, and interests [3], [4], [5].

Giving shape to discourses, **narratives** are persistent stories formed by (re) tellings that express shared meanings [3]. They are linkages and spoken expression of ideas 'into explanation and persuasion' [8, p. 72]. Narratives have been identified as a 'key political strategy to argue for empowering institutional reforms' [9, p. 1031]. Narratives can be mobilised in transformation processes to: (i) change frames by being a counter-narrative to hegemony; (ii) form a collective identity via shared purpose and vision; and (iii) guide action. These 'narratives of change' may eventually legitimise ideas, persuade actors, and transform practices [7].

The narratives of food sovereignty, *Buen Vivir*, and the new plural left will be analysed here using the framework developed by Wittmayer and colleagues [7], which focuses on three interrelated aspects:

- Content: the rationale, relevant actors, and plot of a narrative, including problem framing, visions of the future, roles, actions, and their setting.
- Construction: the ways actors co-create a narrative, the role of technology in its creation/deployment, and its relationship with dominant narratives.
- The role of narrative and the actor's perception of it in the process of social transformation.

This framework is advantageous since it focuses on the process of narrative creation and its role in transformation. Narrative content categories also allow for a clear understanding of change from the perspective of actors, as it establishes the current and desired situations and pathways to achieve change.

The purpose of this section is to delineate the contours of each narrative and make explicit their shared ideational elements. More depth in their analysis and diffusion dynamics is required but exceeds the purpose of this book.

Food sovereignty

The analysis presented below was developed from limited food sovereignty literature [10], [11], [12], [13], [14], [15]. However, it offers a sufficient understanding of ideational elements. Food sovereignty is a contentious concept with multiple definitions. The most widespread was developed in the Nyéléni International Forum on Food Sovereignty (2007), where it was defined as:

> The right of peoples to healthy and culturally appropriate food produced through ecologically sound and sustainable methods, and their right to define their own food and agriculture systems. It puts the aspirations and needs of those who produce, distribute and consume food at the heart of food systems and policies rather than the demands of markets and corporations. It defends the interests and inclusion of the next generation. It offers a strategy to resist and dismantle the current corporate trade and food regime, and directions for food, farming, pastoral and fisheries systems determined by local producers and users. Food sovereignty prioritises local and national economies and markets and empowers peasant and family farmer-driven agriculture, artisanal fishing, pastoralist-led grazing and food production, distribution and consumption based on environmental, social and economic sustainability. Food sovereignty promotes transparent trade that guarantees just incomes to all peoples as well as the rights of consumers to control their food and nutrition. It ensures that the rights to use and manage lands, territories, waters, seeds, livestock and biodiversity are in the hands of those of us who produce food. Food sovereignty implies new social relations free of oppression and inequality between men and women, peoples, racial groups, social and economic classes, and generations [16, p. 1].

Content of narrative

Why does the world have to change? (Rationale)

The neoliberal food system concentrates power in very few transnational corporations while curtailing possibilities for small-scale producers. Food systems are increasingly unsustainable and unhealthy, leading to wicked problems like soil erosion, water pollution, and malnourishment. The current crisis in food systems is caused by industrialism in agriculture, which has led to genetic modification, unhealthy food and eating habits, food insecurity, and a de-peasantisation of rural areas.

Neoliberalism is also a cause, enacted through austerity measures and global trade policies leading to the liberalisation of food markets and financialisation of food. Concentration of power by corporations, biopiracy, and intellectual property rights over genetic material also contribute to the crisis.

The desired future is an overarching vision of feeding the world with small-scale farming. This will be made possible by strengthening peasant models of food production through solidarity-based economic organisation, the promotion of agroecology, and other ecological agricultural alternatives. This vision seeks to return control of local food systems to small-scale producers, promoting autonomy in decision-making and revaluing ancestral and traditional food practices.

Who are the relevant actors? (Actors)

Rural social movements (particularly peasant social movements), small-scale farmers, peasant and family farms, intellectuals, and NGOs are working towards the desired future. It is important to mention La Via Campesina, as it is the main transnational force behind the food sovereignty movement. On the other side, transnational Big Ag, neoliberal states, and supranational organisations regulating trade, food safety, etc. are seen as opposing the desired future.

How is the desired future achieved? (Plot)

The overarching narrative is one of more horizontal relations or collaborations with nature. Opposition to neoliberalism is important, including resisting globalisation, free trade agreements, and challenging international food regulation. The narrative focuses on decentralising power in food systems by maintaining control of food production at local level (territorial self-sufficiency) and transforming state-society relations to influence legislation at local and national level. This includes counteracting or complementing the concept of food security with that of food sovereignty.

Novel state–society relations include more participation in decision-making processes, leading to bottom-up agrarian reform. The objective is to develop pro-peasant legislation, leading to more distributive and procedural justice. State-supported food system transformation would lead to the development of food sovereignty legislation, including socially just allocations of resources (e.g., land), state financing of small-scale solidary food production, and protection for alternative practices.

Decentralisation can be achieved through a plurality of local modes of agriculture, which requires more protection of Indigenous and traditional knowledge, as well as more spaces for agroecology and food production alternatives in university training. Strong rural social movements are also part of this transformation, working on gender equity in food production, establishing food production cooperatives, rescuing traditional knowledge, and protecting and sharing seeds.

Construction of narrative

The narrative positions small-scale, peasant family farming as morally and ecologically superior, challenging industrial agriculture due to its negative social and environmental impacts. The narrative frames capitalism as purely destructive, instead of engaging in destruction–creation dynamics. Due to this framing, popular stories focus on conflicts between peasant and capitalist farming, as well as peasant persistence and resistance in a capitalist epoch. These resistances are not only against capitalism but also against developmentalist states that erode localised agroecological practices.

Actors developed a vision for a sustainable food system that promotes harmonious relations between humans and nature. This opens political spaces of struggle at state level to influence legal frameworks. Food sovereignty challenges narratives of proletarianisation, industrial progress, and globalised free markets as beneficial. Additionally, it questions the benefits of intensive agriculture, agrichemicals, and genetic modification, highlighting their negative impacts on the environment. Critics contend this narrative thread leads to a 'belief in humanity's salvation through small-scale farming' [10, p. 1057].

Role of narrative

Food sovereignty worked as an umbrella term that enabled multiple rural social movements to intersect, despite having different agendas. The concept is sufficiently broad to group diverging interests for rural development, providing a frame where actors can come together to challenge current ways of doing things. Food sovereignty galvanised social movements to actively participate in the development of legislation. In Colombia, Ecuador, and Venezuela, the concept informs the activities of SGNs and other rural social movements (e.g., land rights, dignification of farming). In Ecuador and Venezuela, the concept was enshrined in legislation and has become a national directive. There are multi-scalar efforts to operationalise this narrative through practices enacted by AfroIndopeasant food producers across the region and beyond. Crucially, food sovereignty reframes small-scale and family farming as a futural practice instead of an obstinate remnant of a bygone era. It makes these practices visible as a critical contributor in food systems, particularly in poor postcolonial countries.

Buen Vivir[1]

Buen Vivir is often defined as a way of living in harmony with oneself, community, society, and nature [17], [18], [19]; in short, a plurality of good coexistences [20]. *Buen Vivir* could be understood as a set of ideas that are simultaneously 'a radical critique to Development … and a set of alternative proposals where wellbeing is only possible in socially and ecologically expanded communities' [21, p. 6].

Buen Vivir is a discursive innovation [22] that does not pursue a return to an idealised past [23] but looks to the future while conceptualising and building alternative utopias [21]. Accordingly, it can be understood as a 'process of collective social innovation' [24, p. 152]; a 'conceptual and affective platform, another way to understand the world, or even other cosmovision' [21, p. 9]; an epistemic alternative and possible 'line of flight'[2] from the dominant economic paradigm [25]; or a 'rhetoric-ideational resource that breaks with dominant language and mental models, giving it a subversive political and cultural character facing the advance of a totalising and globalising Eurocentric modernity' [21, p. 15].

Content of narrative

Why does the world have to change? (Rationale)

Humanity is facing multiple crises, particularly environmental, economic, and social. The root of these crises is the colonial matrix of power, the continuation of colonial structures resulting in a racist, industrialist, neoliberal capitalist system that works against nature's health and people's wellbeing. Modernity/Coloniality continues to this day as Development, which maintains and exacerbates international power and economic asymmetries. A *Buen Vivir* future envisions a biocentric post-capitalist world that embraces plurality, where people live in harmony with themselves, others, and nature.

Who are the relevant actors? (Actors)

There is a heavy focus on social movements as agents of change, especially those that group identities marginalised by Coloniality (mainly Indigenous and Afro-Latino populations). Other Latin American social movements, such as peasant, LGBTQ, feminist, ecologist, worker, and student movements, are also working towards this future. More marginally, intellectuals, politicians (mainly from the new plural left), the state, and cooperation agencies are also considered part of the *Buen Vivir* movement. It is important to highlight its understanding of nature as agent, particularly as Pachamama (Andean deity of nature), who played an important role in Ecuadorian transformations.

Actors who oppose the desired *Buen Vivir* future are right-wing politicians, capitalists, extractivist governments in Latin America, and supranational entities like transnational corporations and international financial institutions (e.g., the World Bank). More marginally, there is some resistance from traditional leftist intellectuals and politicians (e.g., from some Marxist strands).

How is the desired future achieved? (Plot)

Central to *Buen Vivir* is a positioning of nature at the centre of human activities. This would entail a widening of participation to include nature in decision-making and phasing down extractivist economies. Participation should also be expanded to include marginalised actors and move towards the establishment of plurinational states that allow multiple modes of decision-making and diverse economies/modes of production. This includes Indigenous self-determination and the empowerment of social movements, embracing plurality and complementarity. Some of these events have taken place over the last 15 years and are ongoing. They occur mostly in Latin America (especially in Ecuador and Bolivia) but are increasingly discussed internationally.

Construction of narrative

This narrative is part of 500 years of Indigenous and Afro-Latino resistance to Coloniality in Latin America. *Buen Vivir* positions itself in opposition to dominant narratives, including linear progress, perpetual economic development, neoliberal capitalism, racism, patriarchalism, nature as standing reserve, and monolithic states.

Indigenous communities and intellectuals challenged notions of Western sustainable development in the 1980s, aligning with a resurgence of Indigenous social movements in the 1990s, and their positioning as influential political actors. In the last decade of the 20th century, the Quichua neologism *Sumak Kawsay* was developed in the Ecuadorian Amazon. The concept spread through social networks, academics, and independent media over 15 years, until it became a central component of the Ecuadorian and Bolivian constitutional processes. This garnered international recognition for the concept, as it continues to transform Latin American society at multiple scales.

Role of narrative

Buen Vivir was able to extend beyond the Ecuadorian Amazon and throughout Latin America to inform multi-scale experiments in many countries. This narrative of change is being used by various Indigenous and non-Indigenous organisations and social movements. It has been used in political projects, leading to very impactful implementations in Ecuador and Bolivia and more measured actions in Colombia. Generally, this narrative is being used by diverse actors across the globe to design and implement alternatives to Western development. *Buen Vivir* works as a flexible vision for the future and a set of loose principles that may be adapted by any community in any context. It positions itself as an alternative, simultaneously criticising dominant narratives ascribed to Modernity/Coloniality while directing the emergence of diverse material micro-utopias.

New plural left

The emergence of a new plural left has been observed in Latin America since the end of the 20th century [26]. It started with a wave of constitutional reforms in the 1990s and continued as many leftist governments gained power from the beginning of the 21st century. The spread of a new socialism that strived to differentiate itself from Marxist projects of the 20th century was dubbed the 'pink wave'.

In Venezuela, this took the form of eco-socialism, in Ecuador it was more akin to a post-neoliberal post-development, and in Colombia it is still consolidating through the first leftist presidency in its history (2022–2026). The new plural left seeks to link social movements, purposefully working towards the inclusion of Indigenous, Afro-Latino, and other marginalised groups to build alternative development regimes. However, these processes generally have not lived up to their promises, being more akin to neo-developmentalist models that end up suppressing other radical alternatives.

Content of narrative

Why does the world have to change? (Rationale)

Capitalism, neoliberalism, and Development have produced a devastating ecological crisis. The model of Development globalised by the United States of America after World War II is unextendible in space and impossible to maintain over time due to planetary biophysical constraints. It also perpetuates colonial asymmetries, maintaining a global division of labour with rich nations at the top. Capitalism is unsustainable; it eats itself as it expands and absorbs any alternative, effectively nullifying other systems. Additionally, the neoliberal state now produces increased levels of inequality and precarisation of life for the global majority. In response, the new plural left aims for a post-capitalist or at least post-neoliberal system that puts diverse social movements at the centre of change production through radical, direct, and deliberative democracy. The state should be reconfigured to serve the interests of the population through dialogue. This future envisions a multi-polar world order where diversity is allowed instead of homogenising difference.

Who are the relevant actors? (Actors)

The actors working towards this future are new plural left politicians, intellectuals, and especially social movements (e.g., the precariat, religious communities, AfroIndopeasant, women). Supranational organisations and economic communities of like-minded nations are also included. Opposing actors include supranational entities like transnational corporations and international financial institutions (e.g., the World Bank), the governments of rich nations, national-level right-wing politicians, and affluent citizens.

How is the desired future achieved? (Plot)

This future is achieved through a widening of participation and radical democracy (social incorporation and direct participation). There is a need to focus on social wellbeing, fraternity, and solidarity while rejecting capitalist individualism. This implies working towards economic alternatives different from communism, including a decentring of the industrial proletariat as an agent of change in favour of social movements representing other marginalised actors.

Economies should transition to be more redistributive, moving away from extractivism. New plural left governments need to strengthen relations with like-minded nations and resist or question rich nations and the globalised institutions that favour them. New plural left projects mobilise decolonial narratives of resistance and use historical symbols. These reconnect people with pre-colonial history and decolonial stories of resistance to, and liberation from, colonial oppression.

The new plural left process can trace its origins to the fall of the Soviet Union and the Latin American constitutional processes of the early 1990s. This repositioned the left, allowing it to come to power in several Latin American nations in the late 1990s and early 2000s. The movement consolidated as a slew of left-wing governments came to power in several countries in the region.

Construction of narrative

This narrative connects to Indigenous and Afro-Latino resistances to colonialism. It frames colonialism as a cultural homogenising project that expands European ways of life globally, while being possible only through the exploitation of non-Europeans. It questions the category of underdevelopment, as it is a consequence of a history of power that has benefited past imperial nations. Crucially, it questions Development as a scientific form of knowledge.

Faced with the failed promises of neoliberalism, increased inequality, and especially the 2008 economic crisis, alternatives were needed to respond in noncolonial, non-Western, post-capitalist terms. This response aims to be nonmonolithic and more a set of plural flexible reactions. As such, this narrative is not openly anti-capitalist, but advocates for post-capitalist or at least post-neoliberal systems. It is a radical departure from Marxist theories of revolution as it decentres the proletariat as main change actor. Linear progress and perpetual economic development are questioned, as these are unsustainable and impossible to reproduce worldwide. Racism and patriarchalism are openly questioned, as is representative democracy.

Role of narrative

The new plural left, or '*movimientista* left', has been positioned to connect diverse social movements with more established political forces. Leftist politicians (both traditional and progressive) have expanded their concerns to include issues of environmentalism, gender, race, and more. This narrative displaces traditional Marxist understandings of a proletarian revolution as a transformative force, favouring the everyday activities of other marginalised communities or networks. Over the last 20-odd years, it has altered state–society relations, bringing the concerns of social movements to the fore in legislation. This novel political space has also allowed for more intersectionality between seemingly disparate social movements.

Shared Transmodern ideational elements

Narrative analysis of food sovereignty, *Buen Vivir*, and the new plural left shows that Transmodern ideals permeate them. Interconnectedness, diversity, and partnership feature heavily, as well as a push for sustainability and an expansion of rights. These narratives also tend to question Modernity and Eurocentricity, rejecting absolutist truth claims and the idea of a linear history. *Buen Vivir* and the new plural left tend to encourage dialogue between diverse ways of being (without excluding the Modern). On the other hand, food sovereignty puts itself in opposition to modern industrial agriculture and aims to feed the world through small-scale farming. Although perhaps not fully Transmodern, it encourages diversity in food production beyond industrialism.

These narratives are glocal, emphasise othered ways of being, and move from geopolitics to biopolitics. They are transnational, being reproduced in cyberspace through information technologies to be adapted in local contexts. These narratives take global concerns and agendas and reframe them in local terms involving diverse actors. Participation is paramount, as direct democracy is favoured, inclusion is demanded, and plurality is encouraged. They push for a post-capitalist or at least post-neoliberal and post-Development and a wider move towards biocentricity.

Are these narratives part of a new era of human consciousness? That claim might be too bold, but they certainly point to the telling of a new shared story of humanity that integrates the pre-Modern, Modern and post-Modern.

Intersecting ideational elements of food sovereignty, *Buen Vivir* and the new plural left provided meaning to the process of Andean seed system transformations. Amalgamations between these narratives produced compelling transformative visions and stories of how to materialise them. However, narratives and the diverse coalitions formed around them were not sufficient to transform whole systems. These narratives were constantly affirmed and reproduced in daily practices. SGNs created alternative micro-institutions through praxis in small-scale experiments that grew to transnational scales. In parallel, state practices opened towards more direct participation influenced by the ideals of a 21st century socialism and the possibility of peace in Colombia. This practice dimension of seed system change is explored in the following section.

Converging practices

Shifting practices and their spread were essential for wider seed system transformations in the region. At its most basic, practices can be understood as 'the consistent rehearsal of the ways in which [people] do things [in a particular setting]' [8, p. 72]. This research argues that both sides of the abyssal line in seed systems (the formal and informalised system) experienced changes in ways of doing things, leading to transformation.

Probing deeper into ways of doing, social practice theory (SPT) shifts attention from individuals and their behaviours to socially shared practices as the site of transformative intervention. In SPT, practices are the source and carrier of meaning, language, and normativity [27]. They are articulations between (i) **meanings** – ideas, aspirations, values, symbols; (ii) **competences** – shared know-hows, practical intelligibility; and (iii) **materials** – technologies, objects, infrastructure [28]. These elements travel constantly but are integrated in localised instances. As such, they are generally stable while practices are always in the process of formation, re-formation, and de-formation [29].

Practices link with each other to form bundles, 'loose-knit patterns based on [practice] co-location and co-existence', and complexes, which are 'stickier and more integrated combinations' [29, p. 17]. For SPT, change is produced through the intentional shaping, directing, or influencing of practices. This change contributes to a sustainable future by stimulating distributed and coordinated governance efforts through practice arrangements [28]. The tasks of practice governance are defined as [29]:

- Affecting the circulation of materials, know-hows, and meanings.
- Undermining support for undesirable practices.
- Facilitating actors' enlistment/defection from (un)desirable practices.
- Shaping networks through which desirable practices propagate or undesirable practices are kept in check.
- Creating conditions to develop and disseminate desirable practices.

Over the last 20 years, SGNs shifted traditional seed care practices through networked interactions. New flows of knowledge and materials made marginalised practices visible to incumbents. On the other side of the abyssal line, states changed their practices to allow inter-epistemic dialogue with the informal(ised) seed system. New modes of participation were implemented and more open approaches to science were encouraged, stimulating the emergence of translation. These practice shifts are framed here as powerful forces in system transformation.

Scaling up seed guardianship

SGNs engaged in social practice governance without naming it as such. These networks intentionally shaped, directed, and influenced seed care and exchange practices, ultimately expanding to interactions with the state and broader society (see Chapter 3). SGNs effectively shaped trans-local and transnational networks through which desirable practices propagate. In this specific case, desired practices are the conservation, propagation, and exchange of native, creole, and landrace seeds according to AfroIndopeasant and agroecological knowledge(s). As SGNs strengthened, network–state and network–society interactions became more relevant, leading to the development of new ways of engaging with these actors.

Meanings ascribed to these practices were explored in Chapter 3. Suffice to say, they are a combination of food sovereignty and *Buen Vivir* ideational elements that reframe native seeds and associated traditional care practices as futural. The main narratives that give meaning to SGN practices are: (i) the opposition between AfroIndopeasant and industrial seeds; (ii) seed guardianship as a practice that produces territorial resilience and autonomy in the face of increased dependence on multinational corporations; (iii) recentring non-Modern agricultural institutions; (iv) pursuing agricultural alternatives to industrialised agriculture; (v) giving dignity and recognition to farming; and (vi) networking as a path to self-sufficiency and food sovereignty.

As part of their networked efforts, SGNs created conditions to develop and disseminate desirable practices and undermined undesirable ones. For these purposes, the networks affected the circulation of materials, know-hows, and meanings. For instance, seed houses materially improve native seed storage, quality, and circulation. On the other hand, spreading meanings of increased dependence on transgenic seeds discourages their use. Additionally, SGN events allowed the free distribution of native seeds (materials), the exchange of practical knowledge about cultivation and storage (know-hows), and the spread of narratives such as *Buen Vivir* or Andean reciprocal nurturing (meanings).

It is important to highlight the role played by information technologies and social media platforms as part of practice governance. There is a correlation between the consolidation of SGNs and the spread of smartphones and social media apps (materials). Platforms like WhatsApp and Facebook are mentioned profusely by guardians as spaces where people connect, knowledge is disseminated, and material exchanges are arranged. In addition to social media, the networks have a very diverse presence online. SGN websites, YouTube videos, online magazines, and other materials circulate beyond geographical borders, producing transnational feedback loops.

Key to this process of transformative practice governance, SGNs did not pursue the dissemination of native seed care *as it was* but sought to transform it. This is evident in the myriad publications (materials) [30], [31], [32] educating guardians (know-how) on appropriate methods and techniques for seed conservation and evaluation, aiming to develop and disseminate participatory guarantee systems [33], or promote peasant feminisms [34].

SGN publications (see [35]) also speak of processes akin to citizen science, where AfroIndopeasant actors use sophisticated genetic tests (materials, know-how) to better understand the quality of their seeds. Additionally, websites (materials) like the Ecuadorian Madre Semilla (www.madresemilla.com) and the Venezuelan El Conuco (www.elconuco.com.ve) aimed to become online exchange platforms and marketplaces for seeds and knowledge. Madre Semilla is a particularly salient case as it hosts massive open online courses, podcasts, databases, publications, and merchandise (know-how, materials) aimed at recruiting new practitioners of Transmodern seed care.

Opening state practices

There was also a noticeable shift in decision-making and research practices of agriculture-related state agencies and high-level governing bodies. Two specific changes are of interest: The broadening of participation through implementation of direct democracy, and an openness to include other forms of knowledge in food system design, regulation, and operation. It is important to acknowledge again the correlation between these phenomena and the expansion of a new plural left, *Buen Vivir*, and food sovereignty narratives (meanings).

State-level direct democracy is fully present in Ecuador and Venezuela but is more marginal in Colombia. In Venezuela, the Bolivarian Revolution and subsequent 1999 constitutional process advocated for street parliamentarism as a way for citizens to become protagonists of democracy. This shift called on civil society organisations to draft and submit laws for review by the National Assembly. These processes occurred due to the creation of legal frameworks and a constant effort to disseminate law-making practices throughout the Venezuelan population. In the case of the 2015 Venezuelan seed law, it is important to highlight that street parliamentarism allowed the participation of historically marginalised peoples from rural areas traditionally excluded from decision-making.

In Ecuador, a more direct and deliberative democracy was instituted by the 2008 *Buen Vivir* constitution. In this case, legislating by popular initiative became the route by which rural social movements, SGNs, and their allies shifted the rules in favour of more Transmodern approaches. The 2008 constitutional process was in itself an open and participatory process, where broad sectors of Ecuadorian society had a voice, particularly those historically marginalised. Beyond the mechanisms of representative democracy, the Ecuadorian constitution enshrined citizen's rights to participation through direct and communitarian democracy. Any citizen group can create, reform, or rescind laws if they have support from 0.25 per cent of people registered to vote. Direct democracy not only manifests through instruments like popular initiatives, but also in the participatory nature of law-making processes like the one by the COPISA.

Regarding research practices, the signing of a peace agreement between the FARC-EP guerrilla organisation and the Colombian state opened new spaces for peasant and family agriculture. This had direct consequences on the inner workings of the Colombian agricultural research authority (AGROSAVIA), which changed its culture to promote second-order research practices and more horizontal dialogues with rural actors. In the case of Ecuador, the Ministry of Agriculture and Livestock (MAG) created the undersecretary of peasant and family agriculture, as well as directorates of agroecological and sustainable production and ancestral knowledge. These instances allowed native seeds to become more visible to state actors and implement actions that encouraged informal(ised) seed practices to consolidate and spread (e.g., through support for native seed fairs). In Venezuela, the 2015 seed law recognised local knowledge-holders (*maestres pueblo*) and created a scientific–peasant alliance that develops participatory research from the knowledge(s) and concerns of marginalised food system actors.

Transmodern practice bundles

The transformation of traditional seed care, policymaking, and research practices had a profound impact on Andean seed systems. Non-Modern meanings and competences gained more weight in the early 21st century due to SGNs, social movements, and state actors coming together through Transmodern discourse coalitions. New materials accompany these shifts as actors mobilise their creative intelligence to implement visions of a post-capitalist post-Development on the way towards *Buen Vivir*. This has effectively created practice combinations that, while still fragile, might point to the establishment of new ways of doing things.

SGNs affected the circulation of materials, know-how, and meanings associated with traditional seed care. This created conditions to develop and disseminate new practices of seed guardianship, while undermining support for transgenic seeds and other dominant food regime practices. In parallel, state practices opened new spaces for dialogue between knowledge(s) to emerge. This shift was achieved in no small part by pressures from below where actors made informal(ised) food systems visible. Additional pressures of decolonial *Buen Vivir*, a new plural left, and food sovereignty aided the Transmodernisation of the state towards broad participation, second-order research, and the recognition of plural ways of understanding food and the world.

Through the simultaneous transformation of state practices and seed guardianship, both sides of the abyssal line became visible to each other and were encouraged to collaborate, producing new articulations. A recognition of reciprocal incompleteness expanded the parameters that define authorship, intelligibility, and purposefulness, effectively opening translational contact zones. This recognition would not have been possible without novel links between materials, know-how, and meanings to produce new ways of doing on both sides. It is still uncertain if these practices will extend over time and create stickier and more integrated combinations. However, it is certain that system rules have changed, opening the way for stronger networks of Transmodern practices.

Changing rules of the game

The cases presented in Chapters 4–6 are only snapshots of ongoing transformation processes. They end in changed rules, understood here as formal and informal conventions that determine appropriate behaviour and are the basis of exchanges between system actors [8]. They exist through use and reproduction in practices and can be regulative (e.g., laws), normative (e.g., values), or cognitive (e.g., belief systems) [36]. Crucially, and connecting to previous sections of this chapter, 'narratives are often used to justify the existence of rules [and] rules often formalise well-established practices' [8, p. 72].

The emergence and consolidation of informal rules within SGNs merit more attention. As this chapter focuses on the emergence of translational contact zones, it covers emerging formal rules resulting from processes of social institutional change negotiated directly by communities of actors [36]. Formal rules in a system are formally constructed and recorded, are enacted by actors through writing and formal interpretation, and are sanctioned by coercive action through formal rewards and punishments [8]. In all three cases reviewed, formal rules were the main arena of contention, as recognition of seed guardianship was seen by SGNs as a necessary condition for the survival of native seeds and AfroIndopeasant ways of life.

The networks dedicated considerable resources to be included in seed legislation. Reviewed change processes can be framed as informal(ised) system actors moving from being rule-takers to rule-makers. It is important to recognise here that 'actors are not [only] passive rule-followers … but active rule users and makers. [They] use rules to interpret the world, make sense, and come to decisions' [36, p. 403]. Despite not being passive rule-followers, AfroIndopeasant actors were historically marginalised by formal seed system rules to the point of almost becoming invisible.

Many authors contend that seed system rules and food systems more broadly tend towards rendering AfroIndopeasant seed practices illegal [37], [38], [39], [40]. This work deviates from those interpretations to argue that something different but equally damaging has happened in seed systems – relegating native seeds and associated practices to what Santos defined as 'the territory of the lawless, the a-legal, the nonlegal', a manifestation of the abyssal line as it tends to 'eliminate whatever realities are on the other side' [1, Ch. 4]. Others have argued that this exclusionary legal regime constitutes a legal liminal space. Liminality emerges from anthropocentric and capital-centric frames that keep seed guardianship in the margins, as it defies classification within those paradigms [41].

This book argues that despite these circumstances, the informal(ised) seed system moved from a liminal/non-legal space to a legal space through the opening of translational contact zones and epistemic justice.

How did informal(ised) system actors escape this a-legal space? Transmodern narratives and practices were crucial. Seen through the lens of new institutionalism, Andean seed system transformations can be defined as processes of institutional bricolage, or the patching together of diverse institutional elements (narratives–practices–rules) in the elaboration of new rules. In these processes, institutional entrepreneurs (SGNs and their allies) engaged in intentional design to adapt 'the rules of the game … meet the demands of uncertain and changing environments, and … further their own interests or values' [8, p. 70].

The formation of Transmodern discourse coalitions while SGN and state practices converged aided seed guardians in becoming rule-makers. As pointed out by Venezuelan scholars, it might also have to do with a widespread irreverence of the population towards the law and the 'hackability' of political institutions enabled by direct democracy scaffoldings [42].

Marginalised food system actors were never passive rule-followers as they actively resisted formal rules while coordinating localised informal rules. As described in the previous chapters, SGNs complemented their acts of resistance by mobilising creative intelligence to propose novel transformation pathways. This opened translational contact zones where reciprocal incompleteness is recognised, eventually changing the rules of the game.

As a result of the involvement of informal(ised) actors and their allies, Colombia created the Resolution 464 of Peasant, Family, and Community Agriculture (2017), policy guidelines for agroecological agriculture, and an ongoing co-visioning process to design a national seed system. In Ecuador, examples include the *Buen Vivir* constitution (2008), the Organic Law of Food Sovereignty (2010), the COPISA (2012), and the Organic Law of Agrobiodiversity, Seeds, and Promotion of Agriculture (2018). In Venezuela, examples include the 2015 seed law, its directive of instituting the Popular Council for the Safeguarding and Protection of Local, Peasant, Indigenous and Afro-descendant Seed, and a Scientific-Peasant Alliance that promotes horizontal inter-epistemic dialogues with local knowledge-holders.

As explained in previous chapters, these new laws and emerging institutions are not a panacea that will solve all food system injustices. However, they are a step towards a future culture that assumes the best of the pre-Modern, Modern, and post-Modern to produce a pluriversal utopia through horizontal intercultural dialogue. These laws are part of a retelling of human history from those who have been othered by Coloniality, especially by SGNs that construct their own visions of the future and meta-narratives of social transformation. Crucially, it is a pathway towards food system justice that starts from a commitment to epistemic justice – a recognition of Coloniality as a structuring force and interculturality as a pathway for the recognition, validation, and inclusion of alternative historically marginalised ways of being – a movement from concealing, ignoring, and eroding the non-Modern, to a solidarity-based epistemology, where mutual education and critique between knowledge(s) is emphasised while prioritising those least advantaged [43].

Conclusion

Analysing Andean seed system change over the last 25 years yielded three main sites of transformation: narratives, practices, and rules. This aligns with new institutionalism theory, which understands institutions as 'stable, recurring patterns of behaviour' [8, p. 55], articulating narratives, practices, and rules to determine appropriate behaviour and otherwise constrain it.

In this theory, institutional change happens when the costs of maintaining an institutional path are greater than the costs of change [8]. As non-Modern seed practices, and the global environmental and social crises became more visible, the cost of change seemed to be reduced compared to maintaining Modern/Colonial trajectories. As mentioned in Part 1 of this book, this responds to the dynamics of the third food regime where extractive and regenerative food systems are at odds while producing food futures.

Regarding narratives, overlapping discourse coalitions were formed by connecting three narratives of change: *Buen Vivir*, the new plural left, and food sovereignty. The alternative visions proposed by these narratives were a strong attractor in the face of multiple environmental and social crises. Shared ideational elements allowed a variety of actors to join forces and mobilise change in food systems and beyond. The informal(ised) seed system became visible to incumbents and common terms facilitated translation between the Modern and non-Modern across the abyssal line. Transmodern narratives challenged Modern conceptions of the world while making the non-Modern legible to incumbent actors in seed systems.

Regarding practices, there was a convergence between SGNs and state practices across the abyssal line. SGNs directed an effort of social practice governance promoting the circulation of new meanings, materials, and competences. This allowed traditional seed care to become part of a more robust practice network capable of recruiting new practitioners while becoming visible to incumbent actors. In parallel, state practices shifted by instituting direct and deliberative democracy and opening spaces for inclusive second-order research. This enabled the emergence of translational contact zones, encouraging inter-epistemic dialogues through novel practices.

Finally, formal rules were the main arena for contention as SGNs strived for the recognition of informal(ised) seed practices. In all three cases, formal rules have changed and keep changing through dialogue, creative tension, and antagonism across the abyssal line. The current state of seed legislation in Colombia, Ecuador, and Venezuela evidences a shift towards distributional, procedural, and recognitive justice (expanded in Chapter 9). Moreover, these transformations point to the potential institutionalisation of a Transmodern seed system that actively encourages non-Modern food practices.

In Ecuador and Venezuela, the state played a preponderant role in legislative change. The constitutional processes in both countries opened a nationwide, multi-system translational contact zone where historically marginalised actors had equal voice to that of traditional political elites. It is important to emphasise that these openings were always produced by social movements (particularly rural) by virtue of carrying the constitutionalist presidents to power. In the case of Ecuador, the previous narrative opening of *Buen Vivir* must be highlighted, as it served as the guiding principle for the 2008 constitution. In the case of Colombia, multiple pressures are bringing both sides of the abyssal line closer (e.g., the FARC-EP peace accords). However, there is still no resolution for seed systems, as negotiations between actors are still ongoing.

To conclude, 21st century institutional change in Colombia, Ecuador, and Venezuela has featured Transmodern narratives, practices, and rules. This claim responds to: (i) an evident push to recentre non-Modern ways of being as valid alternatives responding to the current crises; (ii) more transcultural and biocentric futures being proposed and mobilised connecting diverse actors to alternatives; and (iii) participative processes and second-order research being encouraged to produce Transmodern devices through intercultural dialogue.

Crucial to this Transmodern turn and the opening of translational contact zones, epistemic justice is emerging, i.e., an acknowledgement of Coloniality as a structuring force, and interculturality as a pathway for the recognition, validation, and inclusion of alternative, historically marginalised ways of being and knowing – a movement from concealing, ignoring, and eroding the non-Modern, to a solidarity-based epistemology where mutual education and critique between knowledge(s) is emphasised while prioritising those least advantaged [43].

There is sufficient evidence to claim that translational contact zones emerged and a Transmodern transition has been kickstarted. Processes of seed system change over the last 25 years show varied social sub-systems interacting to produce irreversible patterns of change. Multiple food system actors (SGNs, rural social movements, agribusiness, the state, etc.) have participated in these processes, guided by two opposing visions: the modernisation of agriculture through a green revolution model, and food sovereignty through alternative practices. These processes have been highly antagonistic due to the radical difference of guiding worldviews but are moving towards more creative tension, dialogue, and collaboration.

All three Andean seed system transitions have occurred in similar phases: (i) a decade-long invisible change in the background as narratives of change consolidated and SGNs established as communities of practice; (ii) a point of ignition where the process gains momentum, here interpreted as the 2008 Ecuadorian constitution, the Venezuelan seed law negotiations, and the 2016 signing of the FARC-EP peace agreement in Colombia; and, finally, (iii) structural changes becoming visible as new Transmodern rules and emerging institutions; which move the system towards (iv) a new dynamic state of equilibrium. The following chapter explores this process through the lens of sustainability transitions theory and its process model, the multilevel perspective.

Notes

1 For a deep dive on the concept and its spread see Chapter 6.
2 As in Deleuze and Guattari's concept *ligne de fuite*. For more see [44].

References

[1] B. de S. Santos, *Epistemologies of the South – Justice against epistemicide.* New York: Routledge, 2016.

[2] S. Waddock and P. Kuenkel, 'What gives life to large system change?' *Organ Environ*, 2019, doi:10.1177/1086026619842482.

[3] C. Riedy, 'Discourse coalitions for sustainability transformations: Common ground and conflict beyond neoliberalism', *Curr Opin Environ Sustain*, vol. 45, pp. 100–112, 2020, doi:10.1016/j.cosust.2020.09.014.

[4] J. S. Dryzek, *The politics of the earth: Environmental discourses*, 3rd ed. Oxford: Oxford University Press, 2013.

[5] M. Jernnäs and B. O. Linnér, 'A discursive cartography of nationally determined contributions to the Paris climate agreement', *Global Environmental Change*, vol. 55, February, pp. 73–83, 2019, doi:10.1016/j.gloenvcha.2019.01.006.

[6] S. Waddock, 'Reflections: Intellectual shamans, sensemaking, and memes in large system change', *Journal of Change Management*, vol. 15, no. 4, pp. 259–273, 2015, doi:10.1080/14697017.2015.1031954.

[7] J. M. Wittmayer *et al.*, 'Narratives of change: How social innovation initiatives construct societal transformation', *Futures*, vol. 112, Sep.2019, doi:10.1016/j.futures.2019.06.005.

[8] V. Lowndes, 'Institutionalism', in *Theory and methods in political science*, 4th ed., V. Lowndes, D. Marsh, and G. Stoker, Eds., London: Palgrave, 2018, pp. 54–74.

[9] A. Smith and R. Raven, 'What is protective space? Reconsidering niches in transitions to sustainability', *Res Policy*, vol. 41, no. 6, pp. 1025–1036, Jul.2012, doi:10.1016/j.respol.2011.12.012.

[10] H. Bernstein, 'Food sovereignty via the "peasant way": A sceptical view', *Journal of Peasant Studies*, vol. 41, no. 6, pp. 1031–1063, 2014, doi:10.1080/03066150.2013.852082.

[11] P. Clark, 'Can the state foster food sovereignty? Insights from the case of Ecuador', *Journal of Agrarian Change*, vol. 16, no. 2, pp. 183–205, Apr.2016, doi:10.1111/joac.12094.

[12] D. H. Constance, J. Konefal, and M. Hatanaka, *Contested sustainability discourses in the agrifood system*. New York: Routledge, 2018.

[13] B. McKay, R. Nehring, and M. Walsh-Dilley, 'The "state" of food sovereignty in Latin America: Political projects and alternative pathways in Venezuela, Ecuador and Bolivia', *Journal of Peasant Studies*, vol. 41, no. 6, pp. 1175–1200, Nov.2014, doi:10.1080/03066150.2014.964217.

[14] K. Peña, 'Institutionalizing food sovereignty in Ecuador', in *Food Sovereignty: A Critical Dialogue*, 2013. [Online]. Available: https://www.iss.nl/en/media/51pena2013-1.

[15] H. Wittman, 'From protest to policy: The challenges of institutionalizing food sovereignty', *Canadian Food Studies/La Revue canadienne des études sur l'alimentation*, vol. 2, no. 2, pp. 174–182, Sep.2015, doi:10.15353/cfs-rcea.v2i2.99.

[16] Nyéléni International Forum on Food Sovereignty, 'Declaration of Nyéléni', 2007. Accessed: 3 Jan. 2023. [Online]. Available: https://nyeleni.org/IMG/pdf/DeclNyeleni-en.pdf.

[17] A. Acosta, 'El *Buen Vivir* como alternativa al desarrollo: Algunas reflexiones económicas y no tan económicas', *Politica y Sociedad*, vol. 52, no. 2, pp. 299–330, 2015, doi:10.5209/rev-POSO.2015.v52.n2.45203.

[18] F. Huanacuni Mamani, *Buen Vivir/Vivir Bien Filosofía, políticas, estrategias y experiencias regionales andinas.* Lima: Coordinadora Andina de Organizaciones Indígenas (CAOI), 2010. [Online]. Available: https://www.reflectiongroup.org/stuff/vivir-bien.

[19] C. Walsh, *Interculturalidad, estado, sociedad. Luchas (de)coloniales de nuestra época*. Quito: Ediciones Abya-Yala, 2009.

[20] A. Acosta, '*Buen Vivir*: A perspective for rethinking the world', in *Degrowth in movement(s): Exploring pathways for transformation*, C. Burkhart, N. Treu, and M. Schmelzer, Eds., Winchester: Zero Books , 2021, Ch. 5.

[21] F. García *et al.*, 'El *Buen Vivir* como paradigma societal alternativo', *Economistas sin Fronteras*, 2016.

[22] A. E. Beling, J. Vanhulst, F. Demaria, V. Rabi, A. E. Carballo, and J. Pelenc, 'Discursive synergies for a "great transformation" towards sustainability: Pragmatic contributions to a necessary dialogue between human development, degrowth, and *Buen Vivir*', *Ecological Economics*, vol. 144, March, pp. 304–313, 2018, doi:10.1016/j.ecolecon.2017.08.025.

[23] N. Chassagne, 'Sustaining the "good life": *Buen Vivir* as an alternative to sustainable development', *Community Dev J*, vol. 54, no. 3, pp. 482–500, 2019, doi:10.1093/cdj/bsx062.

[24] P. C. Benalcázar and F. J. U. de la Rosa, 'The *Buen Vivir* postdevelopmentalist paradigm under Ecuador's citizens' revolution governments (2007–2017): An Appraisal', *Lat Am Perspect*, vol. 48, no. 3, pp. 152–171, May2021, doi:10.1177/0094582X211004910.

[25] J. Polo Blanco and E. Piñeiro Aguiar, 'El *Buen Vivir* como discurso contra-hegemónico. Postdesarrollo, indigenismo y naturaleza desde la visión andina', *Mana*, vol. 26, no. 1, pp. 1–31, 2020, doi:10.1590/1678-49442020v26n1a205.

[26] S. Ellner, 'The distinguishing features of Latin America's new left in power: The Chávez, Morales, and Correa governments', *Lat Am Perspect*, vol. 39, no. 1, pp. 96–114, Jan.2012, doi:10.1177/0094582X11425333.

[27] T. R. Schatzki, K. Knorr-Cetina, and E. von Savigny, *The practice turn in contemporary theory.* New York: Routledge, 2001.

[28] Y. Strengers and C. Maller, *Social practices, intervention and sustainability: Beyond behaviour change.* New York: Routledge, 2016.

[29] E. Shove, M. Pantzar, and M. Watson, *The dynamics of social practice: Everyday life and how it changes.* London: Sage Publications, 2012.

[30] Campaña Semillas de Identidad, SWISSAID, Fondation D'enterprise Hérmes, and Corporación Biocomercio Sostenible, 'Guía metodológica – Redes de custodios y guardianes de semillas y casas comunitarias de semillas nativas y criollas', 2017. [Online]. Available: https://semillas.org.co/apc-aa-files/5d99b14191c59782eab3da99d8f95126/guia-redes-de-semillas-2017.pdf.

[31] Grupo Semillas, 'Producción y conservación de semillas nativas y criollas de buena calidad y sanidad', 2018. [Online]. Available: www.semillas.org.co/apc-aa-files/5d99b14191c59782eab3da99d8f95126/cartilla-produccion-de-semillas_web.pdf.

[32] S. Rodríguez Cervantes, H. Picado Cerdas, E. Agüero Coto, D. Muñoz Solano, Red de Semilleros Campesinos, Centro Nacional Especializado en Agricultura Orgánica del Instituto Nacional de Aprendizaje – Costa Rica, and Red de Coordinación en Biodiversidad, 'Guía para semilleros y semilleras', 2017. [Online]. Available: https://cdn.biodiversidadla.org/content/download/141717/1087254/file/Guía para Semilleras y Semilleros.pdf.

[33] T. Aguilar Gómez, A. M. García Hoyos, M. García Álvarez, and Campaña Semillas de Identidad, 'Sistema participativo de garantia de la calidad de semillas criollas y nativas para las casas comunitarias de semillas en Colombia – Guía metodológica', 2018. [Online]. Available: www.semillas.org.co/portal/tecnicas/1.4 Sistemas participativos de garantias/cartilla_spg.pdf.

[34] Fondo Accion, 'Minga de pensamiento Andaquí', 2017. [Online]. Available: https://semillas.org.co/apc-aa-files/353467686e6667686b6c676668f16c6c/minga-de-pensamiento-de-estrategias-de-gnero-71-72-1-7_1.pdf.

[35] M. Pérez *et al.*, Eds., *Semillas del pueblo: Luchas y resistencias para el resguardo y reproducción de la vida*. Caracas: Editorial La Estrella Roja, 2016.

[36] F. W. Geels and J. Schot, 'Typology of sociotechnical transition pathways', *Res Policy*, vol. 36, no. 3, pp. 399–417, Apr.2007, doi:10.1016/j.respol.2007.01.003.

[37] M. Coscione and V. García Pinzón, 'TLCs, paro nacional agrario y movimiento social en Colombia', *Nuevo mundo mundos nuevos*, Sep.2014, doi:10.4000/nuevomundo.67127.

[38] V. García López, O. F. Giraldo, H. Morales, P. M. Rosset, and J. M. Duarte, 'Seed sovereignty and agroecological scaling: Two cases of seed recovery, conservation, and defense in Colombia', *Agroecology and Sustainable Food Systems*, vol. 43, no. 7–8, pp. 827–847, 2019, doi:10.1080/21683565.2019.1578720.

[39] I. Vargas-Chaves, A. Gómez-Rey, and G. A. Rodríguez, 'La invocabilidad de la accion de cumplimiento frente a los compromisos suscritos en el paro nacional agrario de 2013: Un analisis desde la regulacion de las semillas certificadas en Colombia', *Revista Jurídicas*, vol. 15, pp. 28–45, 2018, doi:10.17151/jurid.2018.15.1.3.Recibido.

[40] N. H. Vidal and L. G. Escobar, 'Epistemic and political struggles against the privatization of seeds and collective knowledges', *Revista Colombiana de Antropologia*, vol. 55, no. 2, pp. 39–63, 2019, doi:10.22380/2539472X.798.

[41] L. Ashley, D. M. Rodríguez Herrera, and A. Hummel, 'Seed saving and seed sharing: Occupying a liminal legal space', *Studia Socjologiczne*, vol. 2025, no. 2, pp. 191–216, 2025, doi:10.24425/sts.2025.154902.

[42] A. Felicien *et al.*, 'Exploring the "grey areas" of state-society interaction in food sovereignty construction: The battle for Venezuela's seed law', *Journal of Peasant Studies*, vol. 47, no. 4, pp. 648–673, 2020, doi:10.1080/03066150.2018.1525363.

[43] B. Reiter, Ed., *Constructing the pluriverse*. Durham, NC: Duke University Press, 2018.

[44] G. Deleuze and F. Guattari, *Anti-Oedipus: Capitalism and schizophrenia*. Minneapolis, MN: University of Minnesota Press, 1983.

8 Transmodern transitions to *Buen Vivir* in Andean seed systems

The previous chapter explored narratives, practices, and rules as the pre-eminent sites for transformation in Andean seed systems. This heuristic allowed for an initial understanding of large-scale change but does not sufficiently explain how it happened over time. This chapter frames Andean seed systems as socio-technical systems (STS) to better understand transformation and deploys sustainability transitions (ST) theory and the multi-level perspective (MLP) to analyse it.

The chapter presents the theoretical underpinnings of ST theory and the MLP model, its shortcomings, and how it engages with power. It then analyses the process of change in Andean seed systems using the MLP model, mapping relevant landscape forces (macrotrends), proposing a hybrid Modern and non-Modern seed regime, and framing seed guardian networks (SGNs) as a Transmodern innovative niche.

Sustainability transitions theory

ST seeks to understand how structural change happens in STS and how it can be steered [1]. Transitions are long-term non-linear processes lasting between 25 and 50 years [2] that: (i) alter the ways in which an STS functions; (ii) usually start slowly due to 'multiple sources of inertia'; (iii) are polycentric, as 'multiple actors, factors, temporal and spatial scales are relevant for shaping [their] dynamics'; (iv) can be triggered or emerge from ongoing developments; and (v) are open, path-dependent processes with uncertain outcomes [1, Sec. 2.3]. In transition processes, sustainability is understood as a moving target and change is planned through 'process-based, multi-scale and systemic approaches … guided by … visions' [3, p. 558].

There are four overarching concepts of sustainability transitions [2]:

- Co-evolution: The interaction between societal sub-systems influences individual sub-system dynamics, leading to irreversible patterns of change.
- Co-design and learning: Knowledge is developed in a complex, interactive design process with a range of stakeholders involved through a process of social learning.

DOI: 10.4324/9781003715092-12

- Multi-level perspective (see Figure 8.1): Conceives transitions as interference of processes at three levels – innovative practices (niche experiments), structure (the regime), and long-term exogenous trends (the landscape).
- Multi-phase: The transition over time is a sequence of four alternating phases: (i) pre-development, in which the status quo of the system changes invisibly in the background; (ii) take-off, the point of ignition after which the process of structural change picks up momentum; (iii) acceleration, in which structural changes become visible; and (iv) stabilisation, where a new dynamic state of equilibrium is achieved.

The MLP states that transitions in STS occur as the result of interactions between processes at different levels [1], [4]: a macro-scale landscape (global trends, cultural paradigms, transnational actors, and agreements); a meso-scale regime (dominant structures, culture and practices, existing actors and networks that want to maintain the status quo); and micro-scale niches (the sites for radical innovation, technologies, practices, and actors that are peripheral to the regime). Interactions between these levels (destabilisation and/or alignment) can lead to different processes. Niches can lead to regime

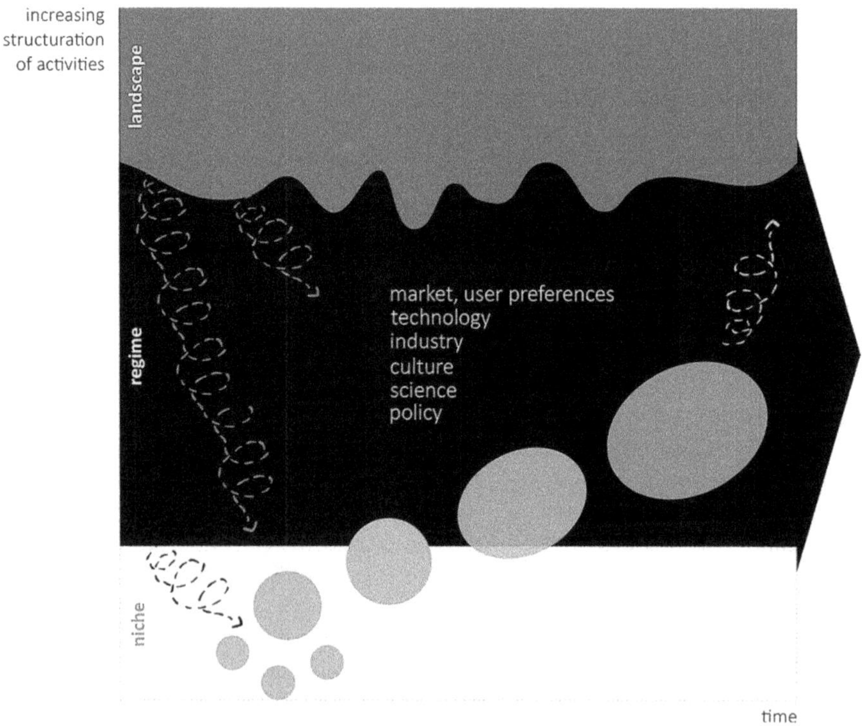

Figure 8.1 The multi-level perspective model

weakening and transformations at the meso-scale. On the other hand, niches can also be absorbed, combined with, or develop a symbiotic relation with the regime [2].

Criticisms to the theory

ST literature has been criticised on several fronts, including its engagement with politics, power and justice, its innovation and technocratic bias, the need for more research in poor nations, and questions about its ethics and inclusion of marginalised actors. Authors have called for deeper engagements with power and social justice issues during and resulting from transition processes [1], as well as more research on the structural power of mindsets, narratives, or worldviews [5]. Additionally, more research in poor countries is needed [1], as existing ST literature seems to not respond well to low-carbon contexts in developing nations [6].

The STS approach might be insufficient to understand transitions in poor nations, as it is hampered by pervasive notions of Euro-Atlantic development trajectories or development through technology transfers from rich nations [7]. This has been associated with initial ST theorisations lacking critical engagement with 'the unsustainability of [development] models … rooted in capitalist modernity' [8, p. 1]. The Modern bias in ST theory also produces 'scarce consideration to normative and ontological pluralism' [8, p. 3], which has been noted elsewhere as a 'lack of attention to social difference (i.e., ethnicity, class, gender)' and ethical aspects in transitions, 'revealing a moral vacuum' [1, p. 16].

ST has also been criticised for not responding properly to agri-food transformations [7], as well as not attending sufficiently to social and organisational innovations common in these systems [9]. Despite all these valid criticisms, this chapter will argue that both ST and the MLP are useful to understand the dynamics of change in low-carbon food systems. However, Coloniality, the abyssal line, and existing Transmodern hybrids must be recognised.

Andean seed systems offer an interesting challenge to ST as seeds themselves are an unsettling technology. In scientific and corporate hands, they are genetic material and sterile commodities subject to intellectual property rights, while in the hands of seed guardians, they can be living beings worthy of spiritual reverence that should circulate freely. Seeds as technology also unsettle knowledge categories. If seeds have been nurtured by humans for millennia, haven't they been technologies since pre-Modern times? If this is the case, then traditional knowledge(s) have been essential in creating and maintaining agrobiodiversity. However, current agricultural framings, suffering from an industrialist bias, see native seeds as naturally occurring materials to be appropriated and commodified by corporate science.

As elaborated in previous chapters, very active and robust AfroIndopeasant and family agricultural institutions have been generally ignored and informalised by incumbent food system actors. In the context of Andean seed systems, science–society interactions for effective change management towards sustainability necessarily include these Transmodern ways of being. The cases analysed show how innovative Transmodern knowledge or othered sciences are mobilised by seed guardian networks (SGNs). SGNs have worked tirelessly to gain recognition and level the playing field so that they can coexist and compete with industrialised agriculture.

The question of power in sustainability transitions

This chapter uses the lens of ST to analyse the transformation processes of three Andean national seed systems elaborating on issues of power [10], [11], [12], [13]. Most recent developments on power have been consolidated in the power in transition framework, where power is understood as 'the (in)capacity of actors to mobilise means[1] to achieve ends' [14, p. 16]. There are three types of power in transitions:

- Reinforcive: The capacity of actors to reinforce and reproduce existing structures and institutions.
- Innovative: The capacity of actors to create new resources.
- Transformative: The capacity of actors to develop (re)new(ed) structures and institutions.[2]

Regimes are spaces of reinforcive power, niches are spaces of innovative power, and niche–regimes are spaces of transformative power. The landscape is a global space that exercises both reinforcive and transformative power. These types of power can engage in different dynamics with one another at the same and/or different levels of the MLP.

Power mobilised through narratives is of particular interest to this book. It can be defined as ideational power, or the capacity of actors to influence the beliefs of others by using ideational elements[3] [15]. Actors exercise power through reasoning and argument to achieve their goals in three ways [15]:

- Power through ideas: Capacity of actors to persuade others to adopt their worldview.
- Power over ideas: Capacity to impose ideas or to resist the inclusion of alternative ideas. This can happen as: (i) powerful actors impose their ideas; (ii) powerless actors shame others into conformity; and (iii) actors resist considering alternative ideas.
- Power in ideas: Capacity of incumbents to control the production of subject positions, and of institutions to constrain consideration of ideas. This can be achieved through depoliticisation of ideas so that they become embedded in society, thus structuring thought at the expense of other ideas.

From this conceptual framework, this chapter seeks to contribute to discussions on the decolonisation of ST literature and its application in food system transformations. These strands of research are young and still in construction but have stablished research agendas [16], [17], [18], [19], [20]. It is important to emphasise that the cases reviewed in this work are critical, extreme, or deviant in ST literature. However, it is precisely their strangeness that allows for reflections on the Modern/Colonial aspects of transitions and how to transcend them.

In the next section, the landscape, regime, and niche levels of Andean seed systems are described, to then focus on niche–regime interactions emphasising issues of power. This book suffers from a niche bias by focusing on SGNs as agents of change. This is common in ST and agri-food transitions literature and there are calls to transcend it [21]. However, a niche focus is important here to highlight the profound differences between worldviews in Transmodern contexts.

A multi-level reading of Andean seed system transitions

Twenty-first century Andean seed system transitions have been long-term and non-linear open path-dependent processes. They emerged from ongoing developments at multiple scales, occurring in hyper-localised seed houses, regional and national territories, and at transnational scales. They moved slowly for the first decade (1991–2001) of a 30-year period and have accelerated massively over the decade between 2015 and 2025. These change processes involved multiple actors (e.g., AfroIndopeasant farmers, NGOs, governments, etc.) in tension over policy, business, science, culture, and technology. This transition altered the ways in which socio-technical seed systems function as new rules emerged that allow for an expansion of native seed use and related knowledge/practices.

Processes reviewed in this book align with the phases of a transition. The year 1992 is framed as the start of the predevelopment phase, where the status quo changed invisibly in the background with the emergence of *Buen Vivir*, the consolidation of neoliberalism, and corresponding resistances in the region (mainly Indigenous and peasant). The take-off phase can be situated between 1999 and 2005, as the Bolivarian Revolution in Venezuela kickstarted the multinational pink tide[4] and SGNs emerged in the region. The transition accelerated in 2015 as the Venezuelan seed law was approved, continuing through 2022 with the co-creation process for the Colombian national seed system. The stabilisation phase is yet to come, since, despite the approval of multiple progressive agricultural laws over the last ten years, marginalised food system actors feel their position is constantly threatened.

Seed systems have transitioned over the last 30 years, picking up momentum since 2015. Although it is hard to predict if this will lead to truly irreversible patterns of change, state–society relations have been fundamentally altered, leading to the creation of novel policy mixes for more Transmodern seed systems. This section will describe this transition using the multi-level perspective (MLP), a process theory developed by ST scholars. Power mobilisation and power relations will be added to the analysis of each level.

Landscape forces

The landscape is an exogenous environment beyond the influence of actors and changes slowly in processes that can take decades. However, rapid external shocks can also be understood as part of this level. The landscape 'highlights the technical, physical and material backdrop that sustains society' [22, p. 403] and enables some actions while hindering others. The landscape exercises both reinforcive and transformative power. Reinforcing landscape developments stabilise the current regime while disruptive developments create impulses for change [22].

Pairs of reinforcive/transformative macro-trends will be described in this section. These trends can also be understood as equivalent to institutionalised and alternative discourses. The landscape level is characterised by meta-discourses or deep-rooted values and assumptions [23]. These landscape-level meta-discourses are 'buil[t] on values and assumptions that are unchallenged and often perceived as common sense in more than one system' [23, p. 1845].

Empirical evidence presented in this book points to strong and fixed abstract ideas challenging established meta-discourses that are shared in various socio-technical systems. As such, the crisis of Modernity, Transmodernity, post-capitalist alternatives, *Buen Vivir*, and a new plural left are included here. Such alternative discourses can arise from long-standing narratives reproduced by specific discursive agents, in this case AfroIndopeasant actors, leftist politicians, etc. (see Chapter 7) [23].

In Chapter 1, Andean seed system transitions were situated in an exogenous environment of two broad macro-trends: the **crisis of Modernity** and the emergence of a new cultural paradigm in **Transmodernity**. Western ways of being and knowing appear to be in crisis. It is becoming increasingly clear that the material intensity of life in rich nations is impossible to sustain and expand. The impacts of the ecological crisis are more evident than ever while social inequality is rampant. Coloniality was established as the structure fuelling these crises.

This book ascribes to the position that there are no Modern solutions to Modern problems, as Modernity itself is the problem. As such, it positions the emergence of Transmodernity as a counter-macro-trend that pursues decoloniality to achieve a more plural way of understanding reality and history from non-Modern perspectives while maintaining active dialogues with Modernity. It is important to emphasise that Transmodernity is an umbrella term that encompasses multiple intellectual projects (e.g., degrowth, *Buen Vivir, Ukama*, post-Development, deep transitions, among others) [24]. It is happening at global scale and can be considered as an external shock.

Neoliberalism has been widely mentioned throughout the book and is here positioned as another macro-trend. The neoliberal narrative can be summarised as a commitment to maintaining economic/financial/profit/shareholder value growth in a context of private ownership of capital, and advocating for individualism, freedom, and competition with minimum state intervention [25], [26].

Neoliberalism is also faced with a counter-macro-trend in a very robust **post-capitalist, post-Development, degrowth, *Swaraj, Buen Vivir*** global movement that attempts to find alternatives to the neoliberal narrative. These movements are also often decolonial and manifest in Andean seed transitions as ***Buen Vivir*** and a **new plural left**. Although still marginal when compared to other globalised trends, their growing influence and impact in the region positions them as macro-level forces.

Buen Vivir posits that humanity is facing multiple crises with their root being the continuation of colonial structures. This racist, industrialist, neoliberal capitalist system works against nature's health and people's wellbeing. *Buen Vivir* strives for a biocentric post-capitalist world that embraces plurality, and where people live in harmony with themselves, others, and nature. *Buen Vivir* went from being a highly localised Indigenous Amazonian concept in the early 1990s to a transnational movement that influenced state-level policy in several countries.

The new plural left or *movimientista* left emerged in Latin America at the end of the last century as a deeper articulation of social movements and the state. This type of left-wing politics frames capitalism as unsustainable, perpetuating colonial divisions of labour, and increasing inequality and precarisation. It aims to develop and implement a post-capitalist or at least post-neoliberal system, with social movements at the centre of change production via radical direct and deliberative democracy. These movements are not labour-focused but expand their concerns to food, water, preservation of nature, gender equality, and other issues.

The **multidimensional crisis of the dominant agri-food system** and its unsustainability is now widely recognised [16], [27], [28], [29]. As part of this macro-trend, some authors argue that there is a new agri-food regime[5] at play [30] and that agroecological transitions are currently underway [9]. In Andean seed systems, this macro-trend is best exemplified by the pervasiveness of food sovereignty narratives contesting power concentration due to neoliberal dynamics.

Food sovereignty envisions a world fed by small-scale farmers through a strengthening of peasant production models and solidarity-based economies. This shift includes a promotion of agroecology and other ecological agricultural alternatives, as well as a revaluing of traditional food practices and increased control of local food systems by small-scale producers promoting autonomy in decision-making. It is also important to mention that the United Nations declared 2014 as the international year of family agriculture, leading to increased research and policy shifts in favour of these agri-food systems. Furthermore, the period between 2019 and 2028 has been declared as the United Nations decade of family farming [31].

It is also important to mention the macro-trend of **global digitalisation and spread of mobile devices**. It clearly aided in the establishment of SGNs, enabling trans-local linkages, and facilitating learning and activism. SGNs were organised as physical and virtual networks that mobilised alternative narratives to validate their practices and reframe them as futural instead of being obstacles to progress. Niche actors deliberately linked to recruit others, articulate diverse causes, and eventually open decision-making spaces to create new system rules.

Transmodern hybrid seed regimes and the abyssal line

The field of sustainability transitions (ST) defines the regime as a dominant set of structure, culture, and practices [2], [32]. This interpretive analytical concept [4] has been used in transition studies and transition management processes at diverse scales. A regime is composed of six mutually reinforcing processes [33]:

- Knowledge, capabilities, and employment of various actors relevant to the maintenance of existing systems.
- Technical infrastructures and institutions that have developed over time to service those systems.
- Economies of scale and markets of incumbent systems.
- The social significance of these systems and their links to political power.
- Mutually reliant clusters of technologies used by these systems.
- Everyday practices and lifestyle values that have come to rely on these systems.

Regimes are not monolithic, as they are often sets of coexisting structures responding to different logics and expectations [27]. Homogeneous conceptualisations of regimes may not be generally applicable in developing nations, where they have a high degree of non-uniformity [33]. ST literature in developing countries provides apt descriptions of corresponding regimes, generally characterised by having a 'weaker state apparatus, less efficient bureaucracies, higher levels of political and economic instability, less transparency in legal proceedings and enforcement of legal frameworks, and relatively high levels of economic and social inequality' [34, p. 199]. These regimes are determined by struggles over limited resources, being composed mainly by primary sector activities with low levels of industrialisation and often relying on foreign resources.

These types of regimes are also described as highly heterogeneous, hybrid, or multiple, often using the formal and informal dipole to account for difference. This multiplicity and the formal/informal institutional divide are conducive to non-egalitarian regimes where distrust is widespread. In developing societies, informal institutions play a bigger role, as exemplified by most people working in informal economies or institutional reliance on clientelism. In these regimes, informal governance is usually 'strong, pervasive and highly resilient, especially in the face of systemic weaknesses of the formal governance regime' [34, p. 200].

Andean seed regimes as Transmodern hybrids

Agri-food transitions literature often mobilises the regime concept for case analysis but seldomly defines it [21]. This book described Andean seed system regimes as overlapping hybrids of formal and informal(ised) sub-systems, each with its own seed types, practices, associated knowledge, values, (in)formal rules, markets, and technologies (see Chapter 1). While there are few instances of collaboration, both sub-systems are mutually dependent, as, for instance, formal systems use native germplasm as inputs while informal(ised) systems respond to formal rules.

Similar processes of hybridisation have been described elsewhere as taking place in the acknowledgement of diversity and divergence, everyday exchanges, knowledge production, exchange and circulation, and rules of the game and institutions [35]. The resulting hybrids are consistent with descriptions of food systems as complex, multidimensional, diverse, and with different levels of formality, or as assemblages of parallel systems serving different producers and consumers [36]. These descriptions also align with reported difficulties establishing 'fully coherent regimes or groups of individuals who share expectations, beliefs or behaviour' [7, p. 208] in developing contexts.

The definition of Andean seed system regimes as hybrid stems from the high percentage of national food provision coming from small-scale, family, and peasant agriculture. Family farming[6] plays a crucial role in guaranteeing food security in Latin America [37]. As such, this work deviates from most literature on agri-food transitions by making a distinction between the dominant food regime and the particularities of Andean food regimes.[7] By shifting the lens from global to regional postcolonial/poor contexts, traditional agriculture can also be characterised as bundles of stable and locked-in institutionalised practices responding to informal rules.

THE GLOBAL AGRI-FOOD REGIME AND MODERNITY/COLONIALITY

The dominant global agri-food regime is recognised as heterogeneous [38] but is generally defined as 'neoliberal, productivist, and industrial agribusiness supported by political incentives and focusing on markets' [21, p. 5]. In Chapter 1, this was framed as the formal sub-system, here described as a global regime composed of

> large scale, (inter)national, corporate businesses, … largely oriented to maximizing profit and shareholder value … dominant trade policy focus [ing] on international and national formal markets and value chains, and [a neoliberal] dominant political doctrine, with the state largely facilitating private business and competition in open markets, rather than protecting local economic systems that cannot compete with global players … The existing regime is largely based on mainstream Western scientific knowledge, in which other forms of knowledge are excluded [36, p. 768].

The dominant Western scientific way of knowing directly influences system rules and is connected to the agricultural knowledge system (AKS), or the 'formal set of institutes and actors charged with fostering innovation, and captur[ing] the stable actor networks, which support agricultural innovation and learning' [39, p. 36].

As elaborated in Chapter 2, we live in Transmodern times where Modern and non-Modern worldviews coexist and constantly hybridise. However, the dominant agri-food regime mostly responds to a Modern/Colonial worldview, which is also referred by post-Development scholars as a One-world World, or 'industrialism, capitalism, modernity, (neo)liberalism, anthropocentrism, rationalism, patriarchalism and secularism' [40, p. 14].

This universalising force has its roots in the Enlightenment project and was consolidated by science, law, the market, and the state as civilising forces transforming infantile savages (peripheral communities) to Westernised subjects (labour force/consumers). Growth is its main purpose, always pursuing the opening of new markets for material consumption. This is the space of the individual without society, homo economicus, consumers, transnational corporations, free trade agreements, patents, digital rights management, Eurocentred white superiority, big extraction, productivism, high energy and high-pressure manufacturing, etc. In this view, humanity-as-one or the global village are only possible through Westernisation [41].

THE NON-MODERN MANIFESTING IN ANDEAN AGRI-FOOD REGIMES

AfroIndopeasant, small-scale, and family agriculture has always been an important part of Latin American food systems. This is evidenced by the large number of family agriculture holdings that currently produce almost (and often more than) half of some food in the region. As such, the dominant global agri-food regime hybridises with family agriculture in the Andes. As described in Chapter 1, this sub-system refers to small-scale farmer-led customary farming that is unregulated or loosely regulated, and is dependent on seed saving practices and trust-based community exchanges sustained by family, peasant, Afro-Latino, Indigenous, small- or medium-scale farmers.

This non-Modern side of the regime cannot be defined in simple terms as it is highly diverse. However, some conceptualisations can be mobilised to better understand it. Post-Development studies would define them as place-centred, community-, and culture-based notions of wellbeing that have less material notions of prosperity [41]. These have a lot in common with peri-capitalist spaces, sites where communities thrive in precarity, being simultaneously inside and outside capitalism [42]. spaces that are part of the no man's land left by decades of (mal)Development, which transformed territories globally through the constant scalability of progress [41].

These spaces are important for sustainable transitions because they can produce innovative Transmodern hybrids to challenge developmentalism [43]. Increasing awareness of Western civilisation's material intensity as a cause of climate chaos positions these spaces as important to transcend the crisis. Non-Modern ways of being are on the frontlines of environmentalism, and associated knowledge(s), narratives, and practices are a guiding force for sustainable transitions.

This book problematises framings of all traditional peasant-led agriculture as being innovative only by virtue of being peripheral. On an empirical level, there is a clear mismatch between SGN innovations and established ways of doing of other informal(ised) actors, which seem to also maintain a status quo (See Chapter 3). Innovative socio-technical arrangements can emerge from the non-Modern pole of the hybrid agri-food regime. However, many of these spaces also evidence lock-in either by maintaining customary practices and forms of governance or (mal)adapting to the global agri-food regime.

It is important to re-emphasise that the non-Modern side of the hybrid regime is actively marginalised, meaning that possibilities to express actors' narratives and affirm their alternative life projects are extremely limited [44]. The informal(ised) seed sub-system is still perceived as an obstacle for modernisation, and AfroIndopeasant and family agriculture actors are generally excluded from decision-making. As such, power asymmetries between formal and informal(ised) seed sub-systems persist, especially as the 'corporate food regime pushes farms everywhere to adopt intens[ive] production methods … especially in low-income countries' [30, p. 78].

Non-Modern ways of being can be generally framed as seeing humans as part of nature, spaces of communities, situated culture, plural knowledge(s), Indigenous knowledge(s), solidarity, association, craft, refurbishment, collaborative decision-making and action, subsistence, eco-sufficient modes, agroecology, permaculture, informal economies, black markets, piracy, technological disobedience, *Swaraj*, frugal/grassroots/inclusive innovation, metabolic value, meta-industrial labour, etc. Some of these characteristics are certainly features of AfroIndopeasant and family agriculture. However, this book argues that these ways of doing are not always innovative, as they often aim to preserve tradition and a particular status quo by resisting change or (mal)adapting to the global agri-food regime.

Seeing the abyssal line in the Andean seed system regime

Seed systems were defined in Chapter 1 as intertwining and overlapping formal and informal sub-systems adapting constantly to each other's pressures. This book probed the formal/informal divide using a decolonial lens, which uncovers the colonial matrix of power as a structuring force for regime heterogeneity in poor nations. It has been argued that the regime is a hybrid between a global industrial agri-food regime and small-scale AfroIndopeasant or family farming, particularly in postcolonial nations where regime variability stems from their position in the colonial matrix of power. This does not exclude framings of regimes in developing nations as being politically and economically unstable, unequal, corrupt, prone to cronyism, and other issues.

Rather, it is an attempt to explain the deep causes of those characteristics to attain a different perspective on what are arguably manifestations of more complex phenomena.

Andean seed regimes are composed by Modern and non-Modern (Transmodern) hybrids that are always in flux (See Figure 8.2). Each side has distinct discourses, practices, institutions, rules, networks, and organisations, and each maintains their own version of a constantly shifting status quo. As evidenced in this book, the Modern core mobilises technoscience and law to expand and promote industrial seed production; by contrast, informal(ised) non-Modern system actors maintain their knowledge(s) and practices, often resisting legislative and other changes from both the core and innovative Transmodern initiatives (e.g., SGNs). Seed regimes are thus divided by the abyssal line and marked by abyssal thinking, impeding the co-presence of Modern and non-Modern sides and stifling learning in non-colonial terms [45].

Power in the hybrid regime of Andean seed systems

This interpretation of the regime level in Andean seed systems does not seek to underplay the immense power asymmetries between the Modern and non-Modern. Modern/Colonial structures have been naturalised over the last 500

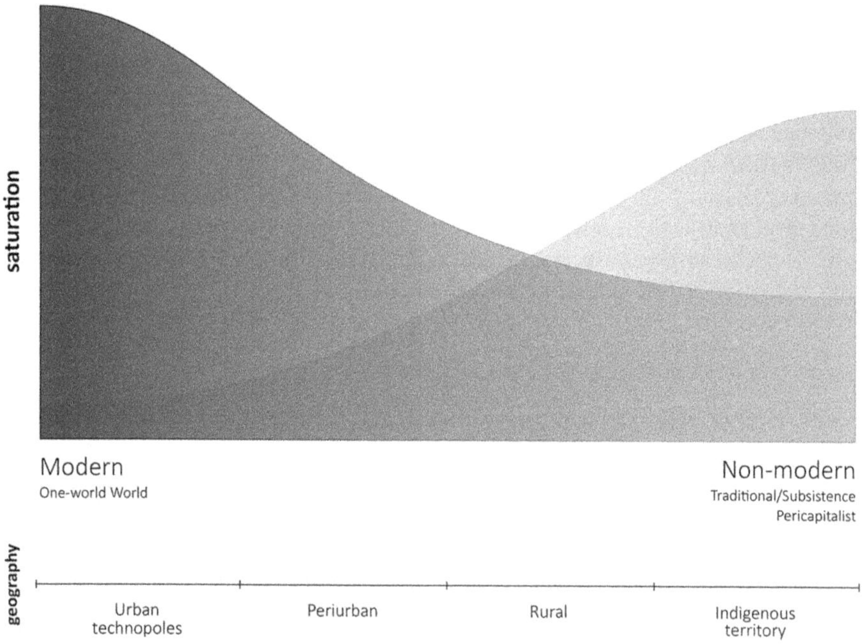

Figure 8.2 A visualisation of hybrid regimes as overlaps between Modern and non-Modern worldviews

years, while their periphery is framed as deviant. The non-Modern has effectively become invisible, inaudible, and illegible to the core, framing Transmodern innovations as utopian eccentricities or threats to societal progress. Ideational power is mobilised by Modern incumbents to control the production of subject positions and institutions, constraining which ideas are considered by actors. This even affects disempowered informal(ised) system actors that resist considering alternative ideas (e.g., Andean reciprocal nurturing) and shame others into conformity.

While informal(ised) system actors are constrained and disempowered by the structural power of Coloniality, formal system actors are empowered and enabled as they mobilise reinforcive power to strengthen the structure. Although formal system actors also mobilise innovative and transformative power by fostering moderate niches (e.g., development of transgenic seeds), they still reinforce Modern/Colonial structures.

Informal(ised) system actors are further disempowered due to other forms of exclusion that marginalise them (e.g., poverty, lack of access to health and education, etc.). Disempowerment also occurs through their disqualification, the destruction and erosion of their social practices, and the invalidation of their knowledge through epistemological racism and control. Despite informal(ised) actors having managed local agrobiodiversity for centuries, they are framed as ignorant and backward because they operate outside of the boundaries of Modern technoscience. This constitutes an epistemic injustice that has harmed generations of peoples since colonial times.

The constantly reinforcing Modern/Colonial power structure influences supranational actors (e.g., the World Bank, transnational seed corporations), and policy instruments (e.g., UPOV, TRIPS). These in turn exert power over governments globally, and influence universities and other formal knowers. At lower levels, large- and small-scale food producers are subjected to this normative power through national seed legislation. This global system of 'power-over' relations remains mostly intact, as it is vast, highly interconnected, and corporate actors can mobilise more resources. For instance, they have a better position in decision-making due to monetary resources. These companies also mobilise human resources by employing people with high levels of education and proficiency in Western codes, facilitating cooperation with influential actors in rule-setting. These actors mobilise mental resources through narratives of capitalist and scientific superiority, marketing, and lobbying to assert their positions. Industrial production of artefactual resources like certified seeds, agrichemicals, and other inputs deployed via vast global supply chains also grants them power (e.g., international development agencies deploying these inputs as part of their programmes).

Supranational agencies, big transnational seed corporations, neoliberal governments, universities, and other formal knowers constantly form reinforcive power coalitions, mobilising mental, artefactual, human, natural, and financial resources. Power of elites, agenda-setting, and preference-shaping manifest in multiple ways in the regime. The agro-industrial bias has been mentioned extensively, as it is pervasive in agenda-setting and preference-shaping by incumbents. The agenda has been set for many decades from Modern/Colonial framings, excluding AfroIndopeasant actors, small-scale family farmers, and agroecological producers. Their issues are intentionally kept off the agenda, as they are seen as obstacles to progress that should industrialise or become part of the agricultural labour force.

Formal and informal(ised) seed systems coexist, compete, and antagonise each other. Tensions within the regime are multifaceted but occur mainly at epistemic level (i.e., technoscientific knowledge vs. traditional knowledge), and in law arising from what is considered legal, illegal, and what is kept in a non-legal space.

This book argues that informal(ised) system actors also mobilise reinforcive power by maintaining tradition or conforming to the global agri-food regime. Be it by virtue of their own autonomous decisions or by the extreme pressures brought upon them by Modernity/Coloniality. As such, informal(ised) system actors 'may also be a source of technological or epistemic lock-in … because of risk aversion or actions of some powerful merchants of doubt' [9, p. 12].

Despite the persistence of Coloniality and its resistance to Transmodern innovation, the non-Modern pole of the regime has seen the emergence of a novel innovative space in seed guardian networks. These networks are peripheral to both Modern and non-Modern worldviews in the hybrid regime. They develop Transmodern knowledge and practices for the preservation and spread of native seeds. As expressed by guardians, they are often seen as eccentrics, radicals, or black sheep by their peers and other social movements. In a similar vein, incumbent Modern actors marginalise them by framing them as ignorant.

This book argues that seed guardians are a particular type of Transmodern knowledge-holder. Through their articulation in trans-local networks, these knowers have effectively created an innovative niche space that has played a key role in Andean seed system transformations.

Seed guardian networks: A Transmodern niche

Niches in ST literature refer to innovative practices that are peripheral to the regime. Actors at this level form constellations that eventually cluster and are empowered to transform the regime [2]. Since most of ST focuses on market-oriented innovations [46], niches tend to be defined as protected spaces (usually by the state and innovation policies), created by technology pioneers and entrepreneurs, and interacting multidimensionally with the regime through various actors [33]. Niches have two interrelated levels: (i) a local niche, grouping local initiatives developing innovations dependent on local conditions; and (ii) a global niche, emerging institutional field, or proto-regime, which refers to a community with shared rules exchanging knowledge and resource flows countervailing the regime [47].

Three simultaneous processes combine for the development of a niche [47]. First is **shielding** path-breaking innovations from selection pressures of the regime related to industry structures, dominant technologies and infrastructures, the established knowledge base, markets and dominant user practices, public policies and political power, and cultural significance. Second is **nurturing** innovations through the building of actor networks, articulating their expectations and those of external actors, and aiding broad learning processes related to multiple socio-technical dimensions. Third is **empowering** innovations so that they become competitive in regime terms, allowing shields to be removed. Empowering can occur to 'fit and conform' (being competitive under regime terms) or 'stretch and transform' (restructuring the selection environment).

Path-breaking innovations emerge in niches but are always connected to existing regimes and landscape developments. Niche actors react to variables at these levels to produce appropriate responses connected to user needs. The emerging innovation is eventually used by small groups of people, leading to the creation of a dedicated learning community that strengthens the practice and stabilises rules. Finally, the innovation is widely diffused and can compete with the regime [48].

Second-order learning and networking (specially outside of niche spaces) have been identified as critical for niche development [38]. Successful niches diffuse socio-technical practices in three ways: (i) replicating small projects within the niche; (ii) recruiting more participants through project scale-up; and (iii) translating niche ideas to incumbents [49].

SGNs can be defined in myriad ways. They are akin to 'alternative food networks' or 'civic food networks' [50]; perhaps a type of 'reflexive localism' since they advance sustainability through the protection of endangered food, biodiversity, and local traditional knowledge [30]. They could also be understood as 'sustainable materialist social movements', which focus on replacing unsustainable practices with alternative ones, producing sustainable material flows and institutions often including the non-human [51]. SGNs can also be framed as alternative milieux in radical politics, new social movements, alternative pathways, or alternative spiritualities and lifestyles [52]. They can be defined as 'frugal innovation, grassroots innovation, and inclusive innovation which utilise local assets and involve Indigenous knowledge systems located outside R&D laboratories' [34, p. 199].

SGNs as grassroots networks are akin to niches 'in that they operate on the margins of conventional agriculture, mainstream public funds, and institutional support frameworks' [38, p. 333]. These grassroots socio-technical arrangements have been generally neglected in ST literature [46], leading to calls for adaptation of ST concepts to these transformative alternatives or finding other conceptual frameworks that might suit them better [53]. For this reason, Chapter 7 explored these spaces through a heuristic of narratives, practices, and rules inspired by new institutionalism. However, keeping in line with the socio-technical focus of this analysis, SGNs will be framed here as grassroots innovations (GIs), or

> innovative networks of activists and organisations that lead bottom-up solutions for sustainable development; solutions that respond to the local situation and the interests and values of the communities involved … grassroots initiatives tend to operate in civil society arenas and involve committed activists who experiment with social innovations as well as using greener technologies and techniques [54, p. 585].

GIs are a bottom-up approach frequently led by partnerships between communities and non-profits [46]. They are socially orientated innovations operating in structures owned by communities, guided by shared values and culture instead of profits, and dependent on voluntary work, grants, and mutual exchange. Regretfully, niche formation and growth in disadvantaged civil society contexts and how GI niches pursue broader influence over regimes has not been sufficiently explored [53], [55]. The next sections attempt to contribute to this deficit by describing SGNs as a GI niche that was eventually empowered to transform seed system rules.

SGNs as grassroots innovative niche spaces

Seed guardian networks (SGNs) were described in detail in Chapter 3 and will be analysed here as niche grassroots innovations. SGNs defend the rights of peoples to produce, store, improve, exchange, and trade seeds according to non-Modern narratives and practices. They can be described as diverse, decentralised, place-based, trans-local, and horizontal coalitions of people and organisations committed to native seed care in Transmodern terms. SGNs have created assemblages of innovative practices peripheral to the hybrid regime and have been empowered to become key players in national seed systems.

The SGN niche emerged from landscape pressures and issues in the hybrid regime. At landscape level, the global spread of neoliberalism resulted in real and perceived threats to the subsistence of marginalised actors mainly from seed legislation. On the other hand, mounting evidence of the detrimental impacts of industrial agriculture led to the consolidation of a global food sovereignty/agroecology movement. Decolonial and *Buen Vivir* narratives were also important as a recent development in a 500-year-long resistance to Coloniality by Indigenous, Afro-Latino, and peasant actors in Latin America. At regime level, the persistence of a formal/informal divide due to the abyssal line perpetuated actor marginalisation. Additionally, political shifts towards a new plural left, and the spread of information and communication technologies (ICT), aided niche emergence.

The SGN niche emerged from civil society (sometimes in collaboration with non-profits), as marginalised pioneers (e.g., AfroIndopeasant actors) Transmodernised their practices while maintaining connections to non-Modern values. These pioneers feel peripheral to both Modern and non-Modern regime worldviews as they are shunned by peers and incumbents alike. The niche is Transmodern as it engages in dialogues with Modernity to transform traditional seed care practices, improving seed technology in non-Western terms by ascribing to alternative visions of what it means to live well (*Buen Vivir*) and produce tasty, healthy, and culturally relevant food (food sovereignty).

This description is consistent with framings of Global South GIs as combining local necessities with post-Development discourses [56]. Effectively, the SGN niche goes beyond the (re)production of innovative practices to also create 'different senses of self, social networks, knowledge production, and post-capitalist economic practices' [55, p. 3]. As such, SGNs are more akin to radical niches, which exercise power challenging dominant macro-trends while strengthening counter-macro-trends [10]. The SGN niche challenges notions of development as technological transfers from rich to poor nations, showing that localised Global South knowledge(s) actively participate in the production of sustainable futures.

Niche formation and development

Regarding niche formation dynamics, novel framings of seed care practices started as local initiatives (e.g., transgenic free territories), which eventually expanded and transformed through the formation of (trans)national networks. SGNs consolidated via emergent bottom-up processes led by committed groups of people (including many volunteers), sometimes aided by national and international non-profits. Accordingly, it is glocal in nature as it links with Indigenous, Afro-Latino, rural, and other social movements across the region, mobilising cyberspace to co-produce and disseminate knowledge and advocate for change. This is consistent with descriptions of other food system GIs as self-organising and emerging from bottom-up processes instead of top-down technology implementations [39]. It also aligns with recognition of steering groups playing a critical role in GI niches [57], and networking as an important step for 'upscaling and transcending the local milieu' [46, p. 977].

According to the theory, GIs need constant support as they face a range of challenges related to formation and diffusion. Maintaining a protected space is especially difficult for GIs [38] due to lack of political capital and funding [55]. GIs also engage in internal ideological conflicts and find it difficult to connect with skilled people [46]. Additionally, their emergence within an unsustainable regime threatens viability and diminishes possibilities to diffuse counter-hegemonic ideas [49].

Although SGNs also face similar challenges, this research argues that all three niche development processes (shielding, nurturing, and empowering) are found in SGN trajectories.

Shielding of this niche was championed by niche actors in what can be framed as a hostile environment. Selection pressures of the global agri-food regime include an overarching industrialist bias, heavy emphasis on scientific knowledge for food production, and increasing control of supply chains by few transnational seed corporations. Additionally, non-Modern seed care practices and SGNs were mostly kept in an a-legal space by incumbents, exposing actors to high uncertainty. On the other hand, the knowledge base, dominant user practices, and culture in non-Modern regime spaces does not fully align with seed guardianship.

These pressures were not enough to undermine the niche as SGNs kept sharing knowledge and recruiting new practitioners across local and national borders through social media and other strategies. The decentralised, trans-local, and virtual nature of these networks allowed them to be flexible and convene/disperse as needed. Geography also played an important role in shielding, as experimentation happened in marginalised spaces (i.e., rural AfroIndopeasant territories). Poor nations are generally unable to assert state power homogeneously, leaving vacant spaces where legislation is difficult to enforce. This allowed niche actors to challenge global seed legislation from their own framings of what is (un)just, perhaps also motivated by an irreverence towards existing laws due to decades of marginalisation.

SGNs nurtured their innovations through network building, articulating expectations, and co-learning. GI growth and development is connected to building social links, increasing the number of network actors, and engaging in peer-to-peer trans-local learning processes that eventually create coalitions [46], [57]. SGNs benefited immensely from the spread of information and communication technologies (ICTs) which allowed them to connect with other outliers that shared their same values and interests. This was complemented with face-to-face local, national, and international events, which allowed greater cohesion.

Alignment of expectations and visions occurred through the organic formation of discourse coalitions around narratives of food sovereignty and *Buen Vivir*. These narratives appear to be sufficiently broad to allow multiple perspectives and agendas to coalesce in defence of native seeds. Discourses are important as they give shape to the socio-political legitimacy of niches [58]. ICTs played a crucial role aligning visions as publications communicated SGN expectations to internal and external audiences in real time. These instances of communication with wider audiences have been shown to be critical for niche emergence and development [39]. Shared frames and discourses allow cross-movement solidarity, which was essential for the transformation of national seed laws [59].

Be it about improving seed quality, legislative changes, or peasant feminisms, co-learning is a fundamental part of the SGN niche. This process is continuous and occurs at trans-local scale, iterating learnings from localised farmers and scaling them to large virtual networks. Learning spaces are opened in face-to-face encounters like seed fairs, but mainly occur in online forums, podcasts, and other digital media. Since SGNs are peripheral to both formal and informal (ised) seed systems, their learning processes can be framed as 'creative destruction' where 'old knowledge and ways of learning are discarded in favour of new approaches or recombined with new ideas or processes … [where] peer-to-peer knowledge dissemination complements a process of dis/aggregation, re/consolidation and de/standardisation' [57, p. 247].

Alternative agri-food niches tend to develop 'their own distinctive knowledge systems (KS) and collaborative modes of learning in the absence of support from the formal Agricultural Knowledge System (AKS)' [39, p. 36]. This seems to be the case with SGNs as they challenge the AKS advocating for inclusion of their plural knowledge(s) through Transmodern dialogues. Despite SGNs having limited access to scientific expertise, they also connected with sympathetic professionals in different spaces who share similar values (e.g., technicians specialised in agroecology). This is consistent with other findings that show that learning from traditional knowledge-holders and dialogues between formal and informal science are imperative for GIs [46], [60].

Niche shields have not been removed, as SGN actors feel constantly threatened by incumbents, be it by virtue of their asymmetrical influence in changing the laws or their appropriation of alternative narratives. However, empowerment has occurred through the (re)positioning of seed guardianship as a practice and seed guardians as a particular type of Transmodern knowledge-holder. The re-framing of their practices as future-making instead of regressive pre-Modern residues has led to better self-esteem and assertiveness of niche actors. This revaluing of their role in society produced the momentum necessary for the niche to influence wider system rules.

Seed systems change in the Andes shows that capacity building is essential for transformation. Much of the energy of SGNs has been directed at empowering marginalised actors by shifting their narratives, creating communities of practice, and showing them how to mobilise politically. The novel concept of seed guardianship allowed for a reframing of seed care practices otherwise understood as obstacles to progress. In this manner, niche innovations become competitive in regime terms as they transcend the abyssal line and offer Transmodern answers to regime issues and landscape pressures. As such, the SGN niche influences the global agri-food regime through a stretch–transform pattern, creating capabilities and attracting resources that 'empower participation in political debates over the future shape of institutions and regime selection pressures' [47, p. 1031].

SGNs actively participated in processes that changed the rules of the game to include non-Modern values, narratives, and practices while pursuing dialogues between traditional knowledge and Modern science. Its activities are directed towards stretching the hybrid regime by transcending the abyssal line. This allows the non-Modern to be seen by the Modern and establish new rules allowing more equal opportunities. However, most GI niches seek to influence the regime to co-inhabit that space without replacing it [53]. This appears to be the case with SGNs, as they do not seem to pursue a dismantlement of industrial agriculture but rather a fairer space for competition.

SGN niche diffusion

Over time, protected GIs contribute to global niches and are taken up by regimes. However, they might need to drastically shift established financial and legal structures to successfully diffuse [46], [53]. GIs easily diffuse geographically since they are low-tech, accessible, and empowering [53]. In addition to geographical spread, diffusion requires GI niches to communicate effectively with external audiences, often aided by niche adaptation or regime shifts [38]. However, translation of GI niche ideas to incumbents is challenging since their values and practices are generally widely divergent from the regime [38], [46], [49], [53].

Intermediaries play a central role in translation by moving from internal networking, systematic shared learning, visioning, and expectation management towards external networking [53]. Due to their horizontal nature, many SGN actors unknowingly served as niche intermediaries by connecting projects, negotiating interests, and establishing priorities to create a common vision [61]. However, special recognition needs to be made to SGN steering groups, which played a key intermediary role between the local and global niche by fostering network cohesion, seeking funding, connecting with allies, and sharing knowledge between (trans)national networks.

The SGN niche is trans-local and transnational, which aligns with other niche processes observed in poor nations [52]. SGN niche innovation is non-Western and non-state-driven, emerging at the grassroots level from social movements and localised (eventually trans-local) communities. The SGN niche developed new Transmodern resources, structures, and institutions that diffused geographically as it consolidated. SGNs are mostly fully horizontal networks that disaggregate and come together emergently and flexibly. As recounted by guardians, one of their intentions is to reach unsupported seed carers in remote places. As such, collaboration happens physically and virtually across national borders and consists mainly of knowledge exchanges between niche participants. The possibility of maintaining constant connection over long distances contributed to vision-building and discourse co-creation, which have been identified as critical for diffusion [62]. Ultimately, niche actors were able to exercise ideational power through narratives and the formation of discourse coalitions, which in this case are bricolages between food sovereignty, *Buen Vivir*, and new plural left framings.

Place matters in GI diffusion [53]. In the case of SGNs, these places can be described as vacant rural spaces where the state has diminished influence.[8] These topographies may not be conducive to the production of large-scale agri-food products (e.g., soy or cotton) and are home to peoples that still hold non-Modern values.

Despite financial limitations and state abandon (or perhaps because of it), SGNs exercised prefigurative power by mobilising multiple resources, e.g., artefactual resources by designing and building seed houses, organising seed exchanges, and producing digital databases; mental resources, exemplified by SGN publications and dialogues between niche actors; and human resources, in a plurality of actors including AfroIndopeasant, agroecological and organic farmers, allied nonprofits, academics, and politicians.

This mobilisation of prefigurative power can also be considered countervailing power as it contests Modern/Colonial framings. Transmodern challenges remove power from formal system actors, allowing for more equal power relations in the system. However, including SGN values, framings, and concerns is no easy task since they are widely divergent from those of the Modern regime. Ideas like seeds as living beings or a commons, reciprocal nurturing practices, biopiracy, transgenic prohibition, and many others unsettle Modern/Colonial frames. This book argues that SGN niche upscaling contributed to the infiltration of these issues in established agendas, opening cracks in the formal system, and resulting in transformed state–society relations.

Despite emerging from a marginalised space and having considerably fewer resources, the SGN niche has been able to diffuse by exercising prefigurative, countervailing, and transformative power in Transmodern terms. This niche was not protected by the state or policy but by internal actors through networking, vision building, and discourse coalitions. There is still a long way to go for the SGN niche to fully challenge the dominant agri-food regime. However, it has transformed system rules to hopefully begin the kick-off phase of a transition.

This shift has been enabled by the opening of translational contact zones by intermediaries allowing innovations to be anchored to the regime. As described in Chapter 2, in these translational contact zones both Modern and non-Modern worldviews engage in dialogue and creative tension to explore configurations of novel structures, institutions, and practices. In the case of seed system transitions, these translational contact zones allowed for more protection of the Transmodern SGN niche through legislation and new state-supported institutions, as described below.

SGN niche interactions with the state

Protection and diffusion of the SGN niche occurred largely due to the activities of network members without state support. AfroIndopeasant and family farmers, non-profits, academics, and other professionals unsettled current ways of doing in the hybrid regime while shielding, nurturing, and empowering the niche. However, GI literature emphasises that support in the wider regulatory and socio-political context is needed for GIs to have broad and sustained influence [53]. Additionally, the state in its regulatory function is often mentioned as a crucial actor for the protection and nurturing of niches [63]. Accordingly, this section describes the shifting relationship between the SGN niche and the state as a crucial part of Andean seed system transitions.

This book adopts a relational understanding of the state as being enmeshed in broader social power relations, including with global-scale actors (e.g., the World Bank) [63]. These relations both constrain and enable sustainability transitions through the formation of coalitions to either maintain incumbency or transform systems. As such, the state is not monolithic and is immersed in 'shifting historical alignments of actors and networks vying for [its] support' [63, p. 76].

It is important to point out that neoliberalism has discredited the role of the state in food provision and that the postcolonial and economically disadvantaged position of the countries studied situates them in asymmetrical global power relations. Despite these barriers and the state's role in maintaining incumbency, its relationship with marginalised food system actors has transformed.

Shifting state–society relations is the product of pressures such as decoloniality, the multidimensional crisis of agri-food systems, and the consolidation of a new plural left. The alignment of these pressures with the innovative alternatives developed by SGNs allowed for their inclusion in food system agendas. The SGN niche took advantage of direct democracy initiatives in Ecuador and Venezuela from the late 2000s. However, it was almost a decade after their foundation (2014–2015) that the networks achieved clear support from state actors. The governments of Colombia, Ecuador, and Venezuela mobilised innovative and transformative power in the early 21st century to Transmodernise seed governance. Legal frameworks further empowered SGNs and other alternatives, leading to reconfigured geometries of power.

This section focuses on 'conflicted political fields … in which coalitions mobilise to support, block, or modify ST policies' [59, p. 279]. Two power coalitions were identified empirically in the transition process as they appear consistently in the data:

i Transnational seed corporations, their national representatives (i.e., industry guilds), neoliberal government actors, and formal knowledge experts.
ii SGNs, other social movements (e.g., rural, ethnic, gender, student), government allies, non-profits, and alternative knowledge experts (e.g., agroecology).

Echoing other findings [59], key mechanisms of mobilisation in Andean seed transitions included: (i) grassroots mobilisation in which social movements form coalitions in support of ST policies,; and (ii) regime mobilisation by organised agents against ST policies, as incumbents prefer less disruptive transition pathways [64]. Collaboration between industrial seed companies and SGNs is absent in this process, as SGNs see corporations as competition at best and as a threat to survival at worst.

Translational contact zones

Seed systems have been framed as Transmodern hybrids between Modern and non-Modern worldviews, where technoscientific and traditional knowledge collide. A decolonial lens reveals that the Modern/Colonial worldview channels violence in the form of epistemic control, the disqualification of social agents, and the destruction of social practices.

The established systems of Modernity/Coloniality, the global agri-food regime, and the agricultural knowledge system have naturalised technoscientific, industrialist, and capitalist food production and control the production of subject positions accordingly. Their structuring reinforcive power comes at the expense of plural alternative ideas of food production and exchange. Academics, technicians, politicians, and other hegemonic actors impose these ideas over system actors and resist the inclusion of alternatives. Additionally, powerless actors like family farmers shame seed guardians into conformity by framing their practices as eccentricities. Despite these barriers, SGNs have been able to persuade others to adopt their worldview by mobilising power through ideas.

In the case of early 21st century Andean seed systems, pressures on the regime facilitated the creation of translational contact zones where both sides of the abyssal line can engage in creative tension. This is a result of translation, a polycentric living process where reciprocal incompleteness is recognised, allowing the expansion of the parameters that define authorship, intelligibility, and purposefulness [45]. The concept of translational contact zones has been selected here as it attests to the massive divergence between Modern and non-Modern knowledge(s). However, ST literature has also theorised similar spaces as user assemblages or interaction arenas [61]. ST scholars have also mobilised literature on knowledge in organisations to better understand the absence of common knowledge at the boundaries between different knowledge domains. These inter-epistemic exchanges are referred to as boundary processes or boundary work [39].

The opening of multi-actor translational contact zones was essential for Andean seed system transitions. In Colombia, the co-creation of a national seed system was kickstarted by government actors with a systemic approach inclusive of informal(ised) system actors. In Ecuador, the COPISA brought together system actors for a long-term, nationwide co-creation process to develop food system legislation. In Venezuela, SGNs mobilised street parliamentarism to assert their narratives/practices as a response to a proposed seed law. This book has argued that these were no mere diversity, equity, and inclusion initiatives or generic co-creation processes but rather highly contested exchanges articulating diverging worldviews.

In these translational contact zones, marginalised non-Modern ways of knowing were mobilised to transcend the scientific and legal abyssal lines. Consensus and conflict in these spaces have been explored in Chapters 4–6, e.g., the pervasive narrative of native seeds being of low quality, which negates the non-Modern care that goes into them. Despite such narratives still permeating seed systems in the region, active listening has occurred on both sides of the abyssal line, expanding the parameters that define authorship, intelligibility, and purposefulness.

The functioning of these translation processes has not been sufficiently explored. However, it is important to point out that they are not only the result of one or two years of co-creation events, but rather the result of many years of articulating novel narratives leading to the formation of broad discourse coalitions. These narrative fields materialised in technologies and practices that unsettle the established system, making the non-Modern visible and allowing for more horizontal dialogues. As such, ideational power is of great importance in this transition since knowledge is both an object of change and an instrument enabling change.

Translational contact zones allowed for new Transmodern configurations of knowledge and law. In Venezuela the Scientific-Peasant Alliance is being implemented, encouraging dialogues of knowledge(s). In Ecuador, the Ministry of Agriculture created the Undersecretary of Peasant and Family Agriculture, a Directorate of Agroecological and Sustainable Production, and a Directorate of Ancestral Knowledge, agencies that are fostering further dialogues between technoscience and othered knowledge(s). In all three cases reviewed, SGNs and the informal(ised) seed regime are on their way to or have shifted from being kept in a non-legal space to being recognised in policy frameworks.

Despite Transmodern practices having better footing in Andean seed systems, informal(ised) system actors are not fully satisfied with the results of translation. SGNs keep pushing for more space for their practices and narratives in seed systems, resisting narrative appropriation by incumbents, and aim to limit the expansion of Big Ag. On the other hand, formal system actors constantly seek to dismantle the laws in Ecuador and Venezuela to further their industrialist agenda and undermine non-Modern knowledge. This shows the difficulties of achieving Transmodern transitions and the need for constant translation across epistemic, scientific, and legal abyssal lines.

Intermediaries across the abyssal line

Some actors have been instrumental for the creation and maintenance of translational contact zones: SGN steering groups, sympathetic academics and technicians working in agroecology or social sciences, allies in government agencies and other social movements, as well as government actors open to alternatives. In ST literature, these types of actors have been framed as boundary spanners, knowledge or innovation brokers, systemic intermediaries, or innovation intermediaries [39], [65]. These transition intermediaries are defined as

> actors and platforms that positively influence sustainability transition processes by linking actors and activities, and their related skills and resources, or by connecting transition visions and demands of networks of actors with existing regimes in order to create momentum for socio-technical system change, to create new collaborations within and across niche technologies, ideas and markets, and to disrupt dominant unsustainable socio-technical configurations [61, p. 1072].

Intermediaries are match-makers forming connections within niches and between niches and regimes. They cross spatial and administrative scales, fostering change and facilitating support in multiple contexts (e.g., policy, market). They are important for networking, forming alliances, negotiating, and developing translation processes addressing knowledge gaps and shifting people's interpretative frameworks. These actors are not neutral as they are immersed in power relations of the system and have strategic objectives connected to the problem space [46], [61], [65].

Several types of intermediaries have been identified, and absence of one type may severely hinder transitions [61]. Many intermediaries are unaware of their intermediation, and although this book has not mapped the 'politics at play', it has framed intermediary normative positioning as Modern and non-Modern. Intermediary types will be connected here with functions and outputs of intermediation [65] to better understand their role in Andean seed system transitions.

First, systemic intermediaries pursue whole system change with an explicit transition agenda operating across all MLP levels. SGN steering groups, intellectuals, and new plural left politicians play this role as they pursue eco-socialism, *Buen Vivir*, food sovereignty, agroecological transitions, and more direct democracy. They engage in boundary work on problem definition, framing existing systems as unsustainable and stimulating reflexive consciousness, leading to new interpretative frames. They also network and mobilise resources to transcend current ways of doing, problematising current deliberative arenas and pushing for more participatory processes. This enhances collective action capabilities, opens space and time for deliberation, builds a shared understanding of the situation, and ultimately changes values and interests while transforming identities and social asymmetries.

Second, regime-based intermediaries are bounded by prevailing socio-technical regime characteristics but are mandated to promote transitions. These are civil servants usually working in agri-food-related agencies which changed their problem frames due to policy implementations. In the case of Colombia, legal frameworks promoting family agriculture and agroecology change state agencies. In Ecuador, the *Buen Vivir* constitution and following legislative boom pursues food sovereignty at national level. In Venezuela, the eco-socialist Bolivarian Revolution mandates more direct democracy and agroecological transitions. Regime-based intermediaries opened new spaces of deliberation to co-create knowledge, networked with a wide range of actors to iterate between goals and means of action, explored controversies, and reduced uncertainties, leading to stabilised agreements and out-scaled/up-scaled outcomes of action.

Third, niche intermediaries advance activities of, and experiment within, a particular niche trying to influence the incumbent system. SGN steering groups and their allies in academia fulfil this role by strengthening seed guardianship conceptually and technically. These actors are central nodes in the networks and have a broad view of their operation. They find problems and mobilise resources to solve them often through knowledge co-production. They convince people through pedagogy not only of engaging in seed guardianship but of the value of feminist framings or non-human inclusion. These actors build shared understandings of seed systems, enrol stakeholders, enhance niche capabilities, and take ownership of action. SGN niche intermediaries iterate constantly between goals and means of action, reduce uncertainties, fears, resistances, and contestations by exploring controversies leading to stabilised agreements and out-scaled/up-scaled outcomes.

Fourth, process intermediaries facilitate niche projects 'often without explicit individual agency or agenda, but in support of context-specific (project-based or spatially located) and/or external (niche, regime) priorities set by other actors' [61, p. 1070]. Individual seed guardians and communities often fulfil this role as their seed houses and guardianship activities are highly localised. They develop these projects using online guidelines and connecting to the knowledge base of the networks. There are also regional steering groups that coordinate projects at local level and articulate in national steering groups to elevate local concerns to the global niche. Some government actors also play a role as process intermediaries as they advance national food sovereignty and inclusion agendas by promoting context-specific priorities. These intermediaries network, engage in problem-finding/solving, co-create knowledge by mobilising ideational resources, and enhance collective action capabilities.

Fifth, user intermediaries translate new technologies to users. Although both regime and niche intermediaries are increasingly convinced of the importance of native seeds, SGN steering groups have played a key role in user intermediation. These actors have been exploring partnerships with local chefs, consumer groups, and other food system actors to make native seeds and food products more widely available and recognised. This has led to the publication of catalogues, cookbooks, podcasts, and other media aimed at convincing people on the value of native seeds and change user values and behaviours.

Intermediaries have been essential for Andean seed system transition processes, especially as it pertains to the opening of translational contact zones. Systemic and regime-based intermediaries played a key role in opening Transmodern inter-epistemic dialogues by problematising current deliberative arenas and pushing for more participatory processes. These intermediaries were able to recognise reciprocal incompleteness, allowing the expansion of parameters that define authorship, intelligibility, and purposefulness. As such, they were able to cross the abyssal line, 'navigating between different social worlds where values are at stake' [39, p. 44].

Transmodern policy mixes emerging

Despite increased research on politics, power, and policy, the primary unit of analysis in ST is still the transition of socio-technical systems and not the mutual transformation of states and societies [66]. Still, this book has given special attention to shifting relations between society and the state, leading to novel policies since 'changes in policies … and changes in socio-technical systems are highly interdependent' [64, p. 1]. As such, this section deals with emerging policy mixes as they may signal the beginning of a take-off phase for Transmodern agri-food transitions in the Andes. This is by no means an exhaustive analysis of policy mixes but a general overview of their significance.

A policy mix generally refers to the combination of several policy instruments and their interaction. These may include policies that support niche innovations but also those that weaken the regime or discontinue technologies [63], [64], [67]. Policy mixes are a combination of three building blocks [67]:

- **Elements** include: (i) a policy strategy which combines quantified ST policy objectives and plans for achieving them; and (ii) instruments that are concrete and interacting tools or governance techniques to address policy problems and achieve objectives. These can be economic, regulation, and information instruments for technology push, demand pull, or systemic concerns, among others.
- **Processes** of policymaking and implementation to solve societal problems where the government is the primary agent making decisions. These include problem identification, agenda setting, policy formulation, legitimisation and adoption, implementation, evaluation or assessment, policy adaptation, succession, and termination.
- **Characteristics** describe the policy mix in terms of consistency (alignment of policy mix elements), coherence (synergic and systematic policy processes), credibility (extent to which the policy mix is believable and reliable), and comprehensiveness (how extensive and exhaustive it is).

Two dimensions have been suggested to analyse the politics of sustainability transition pathways: (i) degree of sustainability or the degree of ambition in the promotion of environmental sustainability by actors/policies; and (ii) degree of disruption or the level of disruptive system change targeted by actors/policies [64].

In Colombia, free trade agreement negotiations with the USA, Canada, and the EU in the early 2000s led to a ratification of UPOV 91 and the creation of Resolution 970 (2010), which established requirements for seed production and commercialisation. This instrument was inconsistent, narrow, and lacked coherence, leading to a lack of credibility by informal(ised) system actors. Neoliberalisation of the food system was also a top-down process implemented without proper consultation with family agriculture actors. This led to widespread resistance to these policies, culminating in the termination of Resolution 970 to be replaced by the more coherent Resolution 3168 (2015). The peace accord between the Colombian government and the FARC-EP guerrilla organisation (2016) prioritised rural reform and connected to narratives of *Buen Vivir*, leading to the creation of Resolution 464 for peasant, family, and community agriculture (2017). This resolution opens more spaces for agroecology, plurality, and inter-epistemic dialogues while hinting at food sovereignty and *Buen Vivir* framings. Although no seed-specific policy instruments have emerged in the last decade, a co-visioning process and policy guidelines for agroecological agriculture have emerged (2018–2025).

In Ecuador, the 2008 *Buen Vivir* constitution transforms the state to pursue post-capitalist alternatives in harmony with nature. Participatory democracy and food sovereignty were pursued as actionable strategies. In 2010, the organic law of food sovereignty was approved to promote agroecological, organic, and sustainable agriculture. This law privileges small producers, promotes the recovery of agrobiodiversity and associated ancestral knowledge, and promotes dialogues between knowledge(s). These policies mandated the creation of the Plurinational and Intercultural Conference of Food Sovereignty (COPISA) (2012). This participatory space allowed multiple food system actors across the country to co-create a food sovereignty legal framework. Crucially, the 2017 Organic Law of Agrobiodiversity, Seeds, and Promotion of Agriculture emerged from COPISA. This law recognises the role of native seeds and associated ancestral knowledge, sustainable agriculture, and peasant families in the Ecuadorian seed system. It provides clear definitions of seed types and practices, offering a coherent description of a hybrid conventional/non-conventional seed system operating under principles of interculturality.

In Venezuela, the 1999 Bolivarian constitution reframed the state as eco-socialist. It included the pursuit of food sovereignty, as well as direct democracy strategies to open participation in policymaking. The creation of the 2015 seed law was a contentious process with three different drafts being produced by the National Assembly, industry actors, and informal(ised) actors. Processes of popular debate or street parliamentarism led to a hybrid seed law that recognises and promotes the formal and informal(ised) seed system. This law is comprehensive and coherent as it defines each system, seed types, and practices in detail. The law proposes the creation of a Popular Council for the Safeguarding and Protection of Local, Peasant, Indigenous and Afro-descendant Seed, and a Scientific-Peasant Alliance to promote knowledge exchanges between technoscience and traditional knowledge(s). Crucially, it recognises local knowledge-holders as *maestres pueblo*, who play an important role in the national food system.

Andean seed system policymaking processes have recognised a hybrid Transmodern regime with highly divergent worldviews, values, narratives, and practices. The process went on to set agendas that transcend the abyssal line and include historically marginalised actors. This was not only a state-led initiative as informal(ised) system actors fought along the way for true inclusion of their agendas. Connecting to new institutionalism, these processes can be framed as institutional bricolage, where niche and regime actors become institutional entrepreneurs and engage in collaborative activities of institutional design [68]. These processes have been aided by translation as described by Santos [45], an irreverence towards the law from informal(ised) actors, and the 'hackability' of state institutions in more direct democracies.

Policy formulation has been generally participatory, leading to increased legitimisation. However, implementation and adoption have been less than stellar due to economic slumps and regime-level tensions over food system visions. Additionally, navigating the need for economic growth and

Transmodern/post-capitalist framings leads to difficulties aligning policy mixes. More powerful ministries handling trade, industry, and finance reinforce competing and less sustainable pathways, effectively undermining transitions [63]. Additionally, the cases show no signs of regime-weakening policies, as they allow the operation of a hybrid seed system. Measures like the prohibition of transgenics have been backtracked or kept in a grey area, mostly keeping the formal seed system intact.

Laws connected to transitions are not neutral as they embody and reinforce class interests [63]. Despite keeping the formal seed system intact, emerging policy mixes in Andean countries opened possibilities for marginalised actors. This book argues that consistent, comprehensive, and coherent Transmodern policy mixes have emerged and are being implemented to gain legitimacy. However, it is important to recognise that they sit alongside countervailing policies that seek to maintain the status quo not only in seed systems but in the wider political economy. Although credibility of these Transmodern policy mixes is still precarious due to the multidimensional crisis Andean countries face, they are ambitious in the promotion of environmental sustainability by pursuing post-capitalist and *Buen Vivir* futures. Despite not disrupting the formal seed system, this book argues that Transmodern policy mixes will be disruptive in the long term as informal(ised) system actors are further empowered, and more knowledge production and exchange are encouraged.

Conclusion

This chapter has presented an analysis of Andean seed system transformations using the theories of sustainability transitions and the multi-level perspective, with special attention given to power. It moved beyond analyses of transitions as technical artefacts moving across space [34] to include worldviews, narratives, practices, power, and knowledge relations. This analysis suffers from niche bias as it was mainly described from the perspective of SGNs. However, a niche focus has allowed for deeper engagements with Coloniality and understanding grassroots initiatives as key innovative players in transitions.

The cases reviewed challenge convergence and catch-up theories in transitions, showing that alternative socio-technical pathways to sustainable futures are emerging from the Global Souths with radically different values from Western Modernity. These Transmodern transitions (here guided by visions of *Buen Vivir*) institutionalise alternatives while destabilising Modern/Colonial structures. The cases also challenge current notions of discursive dynamics in transitions [23], as alternative discourses permeate all levels of the MLP in postcolonial contexts. The formation of multi-actor discourse coalitions allowed for disruptive change by altering the values and assumptions of existing meta-discourses.

Andean seed system transitions in the early 21st century emerged due to landscape pressures, defined here as three opposing pairs of macro-trends (Table 8.1) accompanied by a rise in digital communication technologies.

Table 8.1 Landscape forces in early 21st century Andean seed system transitions

Macro-trend	*Counter-macro-trend*
The crisis of Modernity and the apparent exhaustion of Modern thought and politics in the face of the climate crisis, particularly as it relates to industrial/capitalist growthism.	The emergence of a new epistemic paradigm, or culture here defined as Transmodernity, an umbrella term for diverse alternative intellectual projects.
Neoliberalism.	Pursuit of post-capitalist alternatives.
Crisis of the global agri-food system.	A new emergent agri-food regime at play; and perhaps an agroecological transition towards food sovereignty.

The regime level in Andean seed systems has been framed as a Transmodern hybrid between Modern and non-Modern worldviews. Tensions between these worldviews manifest mainly at epistemic level between technoscientific and traditional knowledge, and in law arising from what is considered legal, illegal, and what is kept in a non-legal space. These hybrids hold disparate (and often opposing) values, narratives, practices, rules, and technological assemblages. This heterogeneity is also divided by the abyssal line, which makes the non-Modern invisible.

The persistence of Coloniality in seed systems leads to massive power asymmetries and perpetuates multiple injustices, as many regime actors are informalised. It is important to recognise these harms and pursue justice in agri-food systems. However, it was argued that informal(ised) actors cannot be framed as innovative only by virtue of being peripheral to Modernity/Coloniality. They may also mobilise reinforcive power and be a source of technological and epistemic lock-in while still holding alternative narratives.

In the early 21st century, seed guardian networks were established as a response to landscape pressures and the injustices of a constantly modernising hybrid regime. This innovative niche connected diverse actors to reframe native seeds as appropriate technologies developed by grassroots knowledge-holders. Aided by non-profits, these knowers organised in local niches by building seed houses and establishing local markets where seeds were gifted, bartered, and sold. After some years, local niches went on to connect trans-locally at national and international scales, forming vast virtual networks aided by the spread of information and communication technologies. The SGN global niche has mobilised: (i) prefigurative power through dialogues between formal and informal science(s); (ii) countervailing power by contesting Modern/Colonial framings; and (iii) transformative power as it was able to change system rules.

The SGN niche was not protected by the state or policy, as it was kept in an a-legal space. Internal actors shielded, nurtured, and empowered the niche by articulating in virtual and physical spaces favouring networking, vision-building, and forming discourse coalitions. SGN steering groups played a key intermediary role between the local and global niche by fostering network cohesion, seeking funding, connecting with allies, and sharing knowledge between (trans)national networks. Niche shields have not been removed, as incumbents have asymmetrical influence in decision-making and tend to appropriate niche narratives. However, state–society relations have transformed allowing for more inclusion of informal(ised) actors in Andean seed systems. This may lead to the formation of a niche–regime and the kick-off phase of the transition.

This book has argued that translational contact zones emerged, allowing for creative tension between the Modern and non-Modern. These processes of translation have led to the creation of Transmodern seed laws that facilitate the implementation of Transmodern institutions. Expansion of participation through direct democracy initiatives was crucial for these processes, in Venezuela via the eco-socialist Bolivarian Revolution, in Ecuador due to the *Buen Vivir* constitution, and in Colombia because of the peace process with the FARC-EP guerrilla organisation. Systemic and regime-based intermediaries played a key role in translation by problematising current deliberative arenas and pushing for more participatory processes. These actors were able to recognise reciprocal incompleteness, allowing the expansion of parameters that define authorship, intelligibility, and purposefulness in seed systems.

Some studies in developing countries have shown that regime instability is a barrier to niche development [34]. The cases reviewed show the opposite, as regime instability favoured the emergence of SGNs. Hybrids between industrial and family agriculture overlap and respond to each other's pressures, resulting in heterogeneous technological assemblages, market preferences, knowledge(s), and rules. Economic pressures and threats to subsistence lead to adaptation and experimentation of the non-Modern in this hybrid space. This is aided by an absentee state producing vacant spaces and power vacuums, allowing non-Modern ways of being to persist. Legislative uncertainty also led to SGN activism pushing for recognition in emerging system rules. It is important to emphasise that this is not an idealisation of instability, as it comes at a high cost for societies in poor nations. Regime instability stems from unfulfilled social contracts and long histories of oppression, but is, in the end, a feature of these societies.

In the cases reviewed the state was both an enabler and an obstacle for niche upscaling. On the one hand, its absence promoted the emergence of solidarity, network formation, and knowledge exchanges in vacant spaces. On the other, it undermined Transmodern peripheral niches due to industrialist biases and a developmentalist frame seeking increased participation in global markets. In this often-contradicting scenario, states mobilised their power to transform their relationship with society, allowing for broader participation. In the case of Ecuador and Venezuela, direct democracy in the form of street parliamentarism or popular initiatives allowed niche actors to better participate in decision-making. Allies in positions of power (mainly through left-wing governments) also aided niche actors by informing on the agendas being set.

The SGN niche is peripheral to and challenges both the Modern/formal and non-Modern/informal(ised) regimes. Niche actors constantly described feelings of being outliers to both sides, as their practices and narratives were seen either as regressive obstacles to progress by incumbents or as eccentricities by other AfroIndopeasant and family farmers. Niche–regime interactions were generally geared towards transcending the abyssal line, allowing other actors to see alternative and valid ways of pursuing seed improvement and exchange. It is important to highlight that niche actors do not advocate for a dismantling of industrial agriculture or formal seed systems, but rather pursue protection from the state via the implementation of egalitarian rules that allow both sub-systems to coexist.

The Transmodern niches described in this chapter are radical innovations that emerged from border thinking, seeking to revitalise subjugated knowledges and reappraise subalternities while allowing integrations with Modernity. These Transmodern niches disrupt the regime by bringing the two poles of the hybrid closer. This process is neither a surrender to the civilising efforts of Modernity/Coloniality, nor an obstinate effort to maintain peripheral positionalities leading to epistemic and technological lock-in. Bringing the Modern and non-Modern closer means mobilising politically to open translational contact zones. In these novel deliberative spaces, process intermediaries expand the parameters that define authorship, intelligibility, and purposefulness of knowledge, innovation, and technology.

The analysis presented in this chapter demonstrates that sustainability transitions theory and the multi-level perspective are useful to understand the dynamics of change in Transmodern socio-technical systems. These theories are applicable to poor postcolonial nations and can aid in the pursuit of deep transitions towards Transmodern futures that are more just and sustainable. However, this requires deeper engagements with the history and epistemic and power dynamics of Coloniality. It also requires seeing the other across the abyssal line to attend to the emergence of alternatives. Designing and embedding sustainability-oriented projects requires a capacity to translate or facilitate intercultural dialogues that move beyond First-World and Third-World fundamentalisms. Process intermediaries in transitions should be well versed in translation dynamics and constantly pursue epistemic justice. For transitions to be fair and truly sustainable, criteria of intelligibility, validity, and purposefulness of knowledge must be expanded to include the non-Modern. This may allow the emergence of a future culture incorporating the best of the pre-Modern, Modern, and post-Modern to produce a pluriversal utopia.

Notes

1 These resources can be mental, human, artefactual, natural, or monetary [11].

2 For instance, new legal structures, physical infrastructure, economic paradigms, or religious ideology.

3 For example, discourse, symbols, myths, narratives, collective memories, stories, frames, norms, grammars, models, and identities.
4 The pink tide refers to a wave of left-wing governments in power in Latin America.
5 To clarify, Levidow's use of the term 'regime' comes from political economy and not sustainability transitions.
6 Family farming has been defined by the FAO (Salcedo and Guzmán) as 'a way of organizing agriculture, livestock, forestry, fishing, aquaculture, and pastoralism, which is managed and operated by a family, and above all, depends predominantly on family labour by both women and men. Family and farm are linked, co-evolve, and combine economic, environmental, social, and cultural functions' [37, p. 26].
7 For a similar distinction in another geographical location see the study on the Hungarian context where Balázs and Pataki write of a 'dual (bipolar) economy' [69, p. 43].
8 In some limited cases SGN spaces have special laws as exemplified by Indigenous territories.

References

[1] J. Köhler *et al.*, 'Modelling sustainability transitions: An assessment of approaches and challenges', *Jasss*, vol. 21, no. 1, 2018, doi:10.18564/jasss.3629.
[2] J. Grin, J. Rotmans, and J. Schot, Eds., *Transitions to sustainable development*. New York: Routledge, 2010.
[3] A. I. Gaziulusoy and H. Brezet, 'Design for system innovations and transitions: A conceptual framework integrating insights from sustainability science and theories of system innovations and transitions', *J Clean Prod*, vol. 108, pp. 558–568, 2015, doi:10.1016/j.jclepro.2015.06.066.
[4] S. Fastenrath and B. Braun, 'Lost in transition? Directions for an economic geography of urban sustainability transitions', *Sustainability (Switzerland)*, vol. 10, no. 7, pp. 11–13, 2018, doi:10.3390/su10072434.
[5] M. P. Schlaile, J. Kask, J. Brewer, K. Bogner, S. Urmetzer, and A. De Witt, 'Proposing a cultural evolutionary perspective for dedicated innovation systems: Bioeconomy transitions and beyond', *Journal of Innovation Economics & Management*, vol. 38, no. 2, pp. 93–118, 2022, doi:10.3917/jie.pr1.0108.
[6] L. P. Pant, B. Adhikari, and K. K. Bhattarai, 'Adaptive transition for transformations to sustainability in developing countries', *Curr Opin Environ Sustain*, vol. 14, pp. 206–212, 2015, doi:10.1016/j.cosust.2015.07.006.
[7] A. J. Wieczorek, 'Sustainability transitions in developing countries: Major insights and their implications for research and policy', *Environ Sci Policy*, vol. 84, September, pp. 204–216, 2018, doi:10.1016/j.envsci.2017.08.008.
[8] G. Feola, O. Koretskaya, and D. Moore, '(Un)making in sustainability transformation beyond capitalism', *Global Environmental Change*, vol. 69, March, p. 102290, 2021, doi:10.1016/j.gloenvcha.2021.102290.
[9] G. Ollivier, D. Magda, A. Mazé, G. Plumecocq, and C. Lamine, 'Agroecological transitions: What can sustainability transition frameworks teach us? An ontological and empirical analysis', *Ecology and Society*, vol. 23, no. 2, 2018, doi:10.5751/ES-09952-230205.
[10] F. Avelino, 'Power in sustainability transitions: Analysing power and (dis) empowerment in transformative change towards sustainability', *Environmental Policy and Governance*, vol. 27, no. 6, 2017, pp. 505–520. doi:10.1002/eet.1777.

[11] F. Avelino and J. Rotmans, 'A dynamic conceptualization of power for sustainability research', *J Clean Prod*, vol. 19, no. 8, pp. 796–804, 2011, doi:10.1016/j.jclepro.2010.11.012.

[12] F. Avelino and J. Rotmans, 'Power in transition: An interdisciplinary framework to study power in relation to structural change', *European Journal of Social Theory*, vol. 12, no. 4, pp. 543–569, 2009, doi:10.1177/1368431009349830.

[13] F. Avelino and J. M. Wittmayer, 'Shifting power relations in sustainability transitions: A multi-actor perspective', *Journal of Environmental Policy and Planning*, vol. 18, no. 5, pp. 628–649, 2016, doi:10.1080/1523908X.2015.1112259.

[14] F. Avelino, 'Theories of power and social change. Power contestations and their implications for research on social change and innovation', *Journal of Political Power*, vol. 14, no. 3, pp. 425–448, 2021, doi:10.1080/2158379X.2021.1875307.

[15] M. B. Carstensen and V. A. Schmidt, 'Power through, over and in ideas: Conceptualizing ideational power in discursive institutionalism', *J Eur Public Policy*, vol. 23, no. 3, pp. 318–337, 2016, doi:10.1080/13501763.2015.1115534.

[16] H. El Bilali, 'Research on agro-food sustainability transitions: A systematic review of research themes and an analysis of research gaps', *Journal of Cleaner Production*, vol. 221, pp. 353–364, doi:10.1016/j.jclepro.2019.02.232.

[17] D. Hopkins, J. Kester, T. Meelen, and T. Schwanen, 'Not more but different: A comment on the transitions research agenda', *Environ Innov Soc Transit*, vol. 34, pp. 4–6, Mar.2020, doi:10.1016/j.eist.2019.11.008.

[18] B. Ghosh, M. Ramos-Mejía, R. C. Machado, S. L. Yuana, and K. Schiller, 'Decolonising transitions in the Global South: Towards more epistemic diversity in transitions research', *Environ Innov Soc Transit*, vol. 41, pp. 106–109, Dec.2021, doi:10.1016/j.eist.2021.10.029.

[19] A. Hebinck *et al.*, 'Beyond food for thought – Directing sustainability transitions research to address fundamental change in agri-food systems', *Environ Innov Soc Transit*, vol. 41, pp. 81–85, Dec.2021, doi:10.1016/j.eist.2021.10.003.

[20] S. Arora and A. Stirling, 'Colonial modernity and sustainability transitions: A conceptualisation in six dimensions', *Environ Innov Soc Transit*, vol. 48, p. 100733, Sep.2023, doi:10.1016/j.eist.2023.100733.

[21] I. C. Melchior and J. Newig, 'Governing transitions towards sustainable agriculture – Taking stock of an emerging field of research', *Sustainability*, vol. 13, no. 2, p. 528, 2021, doi:10.3390/su13020528.

[22] F. W. Geels and J. Schot, 'Typology of sociotechnical transition pathways', *Res Policy*, vol. 36, no. 3, pp. 399–417, Apr.2007, doi:10.1016/j.respol.2007.01.003.

[23] M. C. Simoens, L. Fuenfschilling, and S. Leipold, 'Discursive dynamics and lock-ins in socio-technical systems: An overview and a way forward', *Sustain Sci*, vol. 17, no. 5, pp. 1841–1853, Sep.2022, doi:10.1007/s11625-022-01110-5.

[24] M. Swilling, *The age of sustainability: Just transitions in a complex world.* New York: Routledge, 2020.

[25] C. Riedy, 'Discourse coalitions for sustainability transformations: Common ground and conflict beyond neoliberalism', *Curr Opin Environ Sustain*, vol. 45, pp. 100–112, 2020, doi:10.1016/j.cosust.2020.09.014.

[26] G. Kallis, V. Kostakis, S. Lange, B. Muraca, S. Paulson, and M. Schmelzer, 'Research on degrowth', *Annual Review of Environment and Resources*, vol. 43, pp. 291–316, 2018, doi:10.1146/annurev-environ-102017-025941.

[27] N. Wojtynia, J. van Dijk, M. Derks, P. W. G. Groot Koerkamp, and M. P. Hekkert, 'A new green revolution or agribusiness as usual? Uncovering alignment issues

and potential transition complications in agri-food system transitions', *Agron Sustain Dev*, vol. 41, no. 6, Dec.2021, doi:10.1007/s13593-021-00734-8.

[28] M. C. Rawlinson, 'The climate of food: Justice, truth, and structural change', in *Food, environment, and climate change justice at the intersections*, E. Gilson and S. Kenehan, Eds. Lanham, MD: Rowman & Littlefield International, 2021, pp. 91–118.

[29] J. McGregor, 'The intersection of environmental, climate, and food justice', in *Food, environment, and climate change justice at the intersections*, E. Gilson and S. Kenehan, Eds. Lanham, MD: Rowman & Littlefield International, 2021, pp. 29–46.

[30] L. Levidow, 'European transitions towards a corporate-environmental food regime: Agroecological incorporation or contestation?' *J Rural Stud*, vol. 40, pp. 76–89, 2015, doi:10.1016/j.jrurstud.2015.06.001.

[31] FAO and IFAD, 'Decenio de las Naciones Unidas de la agricultura familiar 2019–2028'. Organización de las Naciones Unidas para la Alimentación y la Agricultura, 2019. Accessed: 10 Feb. 2026. [Online]. Available: https://openknowledge.fao.org/server/api/core/bitstreams/1f6d3ffe-6cb4-4cc0-9462-a60efcdfe584/content.

[32] F. Ceschin, 'How the design of socio-technical experiments can enable radical changes for sustainability', *International Journal of Design*, vol. 8, no. 3, pp. 1–21, 2014.

[33] J. Köhler *et al.*, 'An agenda for sustainability transitions research: State of the art and future directions', *Environ Innov Soc Transit*, vol. 31, January, pp. 1–32, 2019, doi:10.1016/j.eist.2019.01.004.

[34] U. E. Hansen, I. Nygaard, H. Romijn, A. Wieczorek, L. M. Kamp, and L. Klerkx, 'Sustainability transitions in developing countries: Stocktaking, new contributions and a research agenda', *Environ Sci Policy*, vol. 84, December2017, pp. 198–203, 2018, doi:10.1016/j.envsci.2017.11.009.

[35] A. Balanzó-Guzmán and M. Ramos-Mejía, 'Towards epistemic diversity in sustainability transitions: An exploration of hybrid socio-technical systems', *Sustain Sci*, vol. 18, pp. 2511–2531, 2023, doi:10.1007/s11625-023-01370-9.

[36] C. Leeuwis, B. K. Boogaard, and K. Atta-Krah, 'How food systems change (or not): Governance implications for system transformation processes', *Food Secur*, vol. 13, pp. 761–780, 2021, doi:10.1007/s12571-021-01178-4/Published.

[37] S. Salcedo and L. Guzmán, Eds., *Agricultura familiar en América Latina y el Caribe: Recomendaciones de política*, 2014. Accessed: 28 Jan. 2026. [Online]. Available: https://www.fao.org/4/i3788s/i3788s.pdf.

[38] D. Maye, 'Examining innovation for sustainability from the bottom up: An analysis of the permaculture community in England', *Sociol Ruralis*, vol. 58, no. 2, pp. 331–350, Apr.2018, doi:10.1111/soru.12141.

[39] J. Ingram and D. Maye, 'Niche knowledge systems-challenging or invigorating the AKS? An analysis of the Permaculture community in England', in *AgroEcological Transitions: Changes and breakthroughs in the making*, B. Elzen, A. M. Augustyn, M. Barbier, and B. van Mierlo, Eds. Wageningen: Wageningen University & Research, 2017, pp. 35–48.

[40] A. Escobar, 'Transiciones: A space for research and design for transitions to the pluriverse', *Design Philosophy Papers*, vol. 13, no. 1, pp. 13–23, 2015, doi:10.1080/14487136.2015.1085690.

[41] W. Sachs, *The development dictionary: A guide to knowledge as power*, 2nd ed., vol. 18, no. 1. London: Zed Books, 2010. doi:10.2307/1971880.

[42] A. L. Tsing, *The mushroom at the end of the world: On the possibility of life in capitalist ruins*. Princeton, NJ: Princeton University Press, 2015.

[43] J. Martinez-Alier, 'The environmentalism of the poor', *Geoforum*, vol. 54, pp. 239–241, 2014, doi:10.1016/j.geoforum.2013.04.019.

[44] I. Milojević and S. Inayatullah, 'Narrative foresight', *Futures*, vol. 73, pp. 151–162, 2015, doi:10.1016/j.futures.2015.08.007.

[45] B. de S. Santos, *Epistemologies of the South – Justice against epistemicide.* New York: Routledge, 2016.

[46] M. Hossain, 'Grassroots innovation: A systematic review of two decades of research', *Journal of Cleaner Production*, vol. 137, pp. 973–981, 20 Nov.2016, doi:10.1016/j.jclepro.2016.07.140.

[47] A. Smith and R. Raven, 'What is protective space? Reconsidering niches in transitions to sustainability', *Res Policy*, vol. 41, no. 6, pp. 1025–1036, Jul.2012, doi:10.1016/j.respol.2011.12.012.

[48] F.-M. Belz, 'A transition towards sustainability in the Swiss agri-food chain (1970–2000): Using and improving the multi-level perspective', in *System innovation and the transition to sustainability: Theory, evidence and policy*, B. Elzen, F. W. Geels, and K. Green, Eds. Cheltenham: Edward Elgar Publishing, 2004, p. 315. doi:10.4337/9781845423421.00016.

[49] G. Seyfang and A. Haxeltine, 'Growing grassroots innovations: Exploring the role of community-based initiatives in governing sustainable energy transitions', *Environ Plann C Gov Policy*, vol. 30, no. 3, pp. 381–400, Jun.2012, doi:10.1068/c10222.

[50] B. Megyesi, 'Agri-food production in a Hungarian micro-region: Factors influencing farmers' practices', in *AgroEcological Transitions: Changes and breakthroughs in the making*, B. Elzen, A. M. Augustyn, M. Barbier, and B. van Mierlo, Eds. Wageningen: Wageningen University & Research, 2017, pp. 76–87.

[51] D. Schlosberg, 'From postmaterialism to sustainable materialism: The environmental politics of practice-based movements', *Env Polit*, vol. 34, no. 1, pp. 92–112, 2019, doi:10.1080/09644016.2019.1587215.

[52] B. Truffer, J. T. Murphy, and R. Raven, 'The geography of sustainability transitions: Contours of an emerging theme', *Environ Innov Soc Transit*, vol. 17, pp. 63–72, 2015, doi:10.1016/j.eist.2015.07.004.

[53] G. Seyfang and N. Longhurst, 'What influences the diffusion of grassroots innovations for sustainability? Investigating community currency niches', *Technol Anal Strateg Manag*, vol. 28, no. 1, pp. 1–23, Jan.2016, doi:10.1080/09537325.2015.1063603.

[54] G. Seyfang and A. Smith, 'Grassroots innovations for sustainable development: Towards a new research and policy agenda', *Env Polit*, vol. 16, no. 4, pp. 584–603, Aug.2007, doi:10.1080/09644010701419121.

[55] H. L. Lai, 'From protected spaces to hybrid spaces: Mobilizing a place-centered enabling approach for justice-sensitive grassroots innovation studies', *Environ Innov Soc Transit*, vol. 47, Jun.2023, doi:10.1016/j.eist.2023.100726.

[56] A. Smith and G. Seyfang, 'Constructing grassroots innovations for sustainability', *Global Environmental Change*, vol. 23, no. 5, pp. 827–829, Oct.2013, doi:10.1016/j.gloenvcha.2013.07.003.

[57] G. Feola and R. Nunes, 'Success and failure of grassroots innovations for addressing climate change: The case of the transition movement', *Global Environmental Change*, vol. 24, no. 1, pp. 232–250, 2014, doi:10.1016/j.gloenvcha.2013.11.011.

[58] F. W. Geels, M. P. Hekkert, and S. Jacobsson, 'The dynamics of sustainable innovation journeys', *Technol Anal Strateg Manag*, vol. 20, no. 5, pp. 521–536, Sep.2008, doi:10.1080/09537320802292982.

[59] D. J. Hess, 'Sustainability transitions: A political coalition perspective', *Res Policy*, vol. 43, no. 2, pp. 278–283, 2014, doi:10.1016/j.respol.2013.10.008.

[60] A. K. Gupta, 'Innovations for the poor by the poor', *International Journal of Technological Learning, Innovation and Development*, vol. 5, no. 1/2, p. 28, 2012, doi:10.1504/IJTLID.2012.044875.

[61] P. Kivimaa, W. Boon, S. Hyysalo, and L. Klerkx, 'Towards a typology of intermediaries in sustainability transitions: A systematic review and a research agenda', *Res Policy*, vol. 48, no. 4, pp. 1062–1075, May2019, doi:10.1016/j.respol.2018.10.006.

[62] K. Schiller, W. Godek, L. Klerkx, and P. M. Poortvliet, 'Nicaragua's agroecological transition: Transformation or reconfiguration of the agri-food regime?' *Agroecology and Sustainable Food Systems*, vol. 44, no. 5, pp. 611–628, 2020, doi:10.1080/21683565.2019.1667939.

[63] P. Johnstone and P. Newell, 'Sustainability transitions and the state', *Environ Innov Soc Transit*, vol. 27, pp. 72–82, Jun.2018, doi:10.1016/j.eist.2017.10.006.

[64] M. B. Lindberg, J. Markard, and A. D. Andersen, 'Policies, actors and sustainability transition pathways: A study of the EU's energy policy mix', *Res Policy*, vol. 48, no. 10, Dec.2019, doi:10.1016/j.respol.2018.09.003.

[65] P. Steyaert, M. Barbier, A. Loconto, A. Levain, and A. Loconto, 'Role of intermediation in the management of complex socio-technical transitions', in *AgroEcological Transitions: Changes and breakthroughs in the making*, B. Elzen, A. M. Augustyn, M. Barbier, and B. van Mierlo, Eds. Wageningen: Wageningen University & Research, 2017, pp. 257–282.

[66] R. Eckersley, 'Greening states and societies: From transitions to great transformations', *Env Polit*, vol. 30, no. 1–2, pp. 245–265, 2021, doi:10.1080/09644016.2020.1810890.

[67] K. S. Rogge and K. Reichardt, 'Policy mixes for sustainability transitions: An extended concept and framework for analysis', *Res Policy*, vol. 45, no. 8, pp. 1620–1635, Oct.2016, doi:10.1016/j.respol.2016.04.004.

[68] V. Lowndes, 'Institutionalism,' in *Theory and Methods in Political Science*, 4th ed., V. Lowndes, D. Marsh, and G. Stoker, Eds., London: Palgrave, 2018, pp. 54–74.

[69] B. Balázs and G. Pataki, 'Cooperative research for bottom-up food sovereignty and policy change', in *Action Research in Policy Analysis*, K. P. R. Bartels and J. M. Wittmayer, Eds. London: Routledge, 2018, pp. 39–60.

9 21st century Andean seed systems

A just transition

Just transitions and epistemic justice

Latin American seed systems are complex. They touch upon many dimensions (political, economic, socio-technical, cultural, etc.), involve multiple actors with diverging worldviews and future visions, and emerge from interactions at multiple levels and geographic scales. These systems not only determine the livelihoods of food producers and others in the value chain, but also the lives and future of humanity. Engaging with justice in such complex systems is challenging, as justice itself is a contentious concept [1].

This book pursues justice engagements from a structural focus and multiple intersecting conceptualisations. A structural focus demands an understanding of structures as dynamic 'forms of organisation that produce social positions and roles, channelling action, and … circumscribing the possibilities that are open to, accessible by, or closed for people' [2, p. 10]. For McGregor and Kortetmäki, structural injustices emerge when these processes increase the power and opportunities for a group while limiting opportunities to exercise capacities for other large groups. As was explored in Chapter 1, Modernity/Coloniality is understood here as the key unjust structure. Structural injustices unfold from the convergence of actions by multiple actors pursuing their goals in line with specific social-structural processes (i.e., Coloniality), meaning that no particular actor can be blamed for them[1] [3], [4].

Regarding multiple conceptualisations of justice, this work aligns with calls to recognise the shortcomings of justice, and move beyond the justice triad – distributional, recognitive, and procedural justice [1], [5]. Distributional justice refers to the fair allocation of benefits and burdens, recognitive justice refers to valuing diverse perspectives, experiences, and cultures, and procedural justice refers to fairness and transparency in decision-making and rules. Accordingly, food systems are viewed here as intersected at multiple levels by diverse justice dimensions:

DOI: 10.4324/9781003715092-13

- Food justice, or the availability of, and access to, healthy and culturally appropriate food, participation in decision-making, and the connection between food and other social justice issues.
- Environmental justice, or the differential impacts of industry on marginalised communities and their exclusion from decision-making.
- Climate justice, or the unequal distribution of impacts from climate change and related policies [2]. As was mentioned in Chapter 1, climate injustice is increasingly being framed as a tension between rich people with high-carbon lifestyles and masses of poor people who are the least responsible for the climate crisis [6].

Expanding on these initial justice dimensions, our situatedness within transgenerational relations and its ethical implications need greater attention [7]. Such considerations lead to conceptualisations of intergenerational justice or allowing a flourishing existence for future generations. For this purpose, ecological justice is needed, as it is concerned with broader relationships between humans and non-humans, focusing on power and equity [2]. Ecological justice is here understood as particularly transformative for relational patterns established by Coloniality. For sustainable futures to emerge, a worldview based on interrelatedness needs to emerge, since narratives of separation between humans and nature lie at the heart of Modernity/Coloniality.

In all justice dimensions, participation is of great importance. Participative justice deals with the right to participate in decision-making and may be the first step to amend disparities in food systems [2]. This form of justice should balance the need to recognise autonomy and difference while treating all participants similarly. Throughout history, marginalised peoples have been excluded from policymaking processes and have often been subjected to the harm of misrecognition by being homogenised as 'universal persons' [3, p. 40]. Therefore, autonomy and recognition of difference are important. Without autonomy and recognition, pressures over marginalised communities may become invisible and perpetuate dominance under the guise of impartiality [8].

Just transitions

The concept of a just transition originated in labour movements and frontline communities in the 1970s and has been mainstreamed by the United Nations for more than a decade. However, it has been broadly de-historicised and depoliticised, being now promoted by a wide range of actors without any clear definition [6]. Whichever meaning is assigned to the term, scholars argue that unjust transitions are unsustainable, and ignoring justice issues is detrimental to garnering political support for them [9]. As such, calls for just transitions are increasing across all sectors and scales [10].

Three main tensions have been identified in the concept of just transitions [11], between:

- Sustainability and inclusivity: 'Tensions between rapid and bold sustainability action in time-sensitive contexts and procedural justice'.
- Sustainability and recognition: 'Tensions between sustainability performance and recognition of diverse value systems and rights'.
- Sustainability and equity: 'Tensions between achieving sustainability performance and socially and environmentally just outcomes across space' [11, p. 436].

So, what constitutes a just transition? For Ciplet and Harrison, a just transition is 'bold and timely sustainability action' with: (i) 'decision-making that is representative of impacted constituents'; (ii) 'strong recognition of diverse rights and values of marginalised peoples'; and (iii) 'equitable distribution of burdens and harms' [11, pp. 441, 445, 449]. For Williams and Doyon, it is about 'considering the harms and benefits of … transitions, critically examining who is (and who is not) part of these processes, who wins and who loses, and recognising the historical exclusion of peoples and worldviews' [9, p. 144].

For Swilling and colleagues, a just transition is only possible if the overall goal is human wellbeing (income, education, and health) within a sustainable world (decarbonisation, resource efficiency, and ecosystem restoration). For these scholars,

> the structural transformations needed for a just transition will only be achieved when there is a socio-political regime that rests on a strategic coalition of interests that shares this paradigm, uses state institutions to drive a just transition and adopts an appropriate policy and legislative programme that is aligned with the overall goal [12, p. 657].

Drawing on Heyen's [1] justice dimensions of sustainability transitions, a just transition should be mindful of the distribution of causation of environmental problems, environmental goods and bads, and a socio-technical system's benefits and costs. Furthermore, there should be recognition, access to information, decision-making, and legal protection throughout the transition for all involved. Distribution of benefits and burdens from related policies should also be considered.

More radical approximations to just transitions seeking to be inclusive of marginalised peoples call for the inclusion of three justice principles: (i) antisubordination – preventing harm, reparative and restorative mechanisms, and liberating processes; (ii) intersectionality – including multiple concurring identities, and 'subaltern and invisible bodies and practices' [5, p. 1752]; and (iii) relationality – including a politicisation of everyday life, emphasising multiple connections, and developing interventions for other ways of being [5].

Acknowledging past and present injustices is an essential part of transitioning to fairer futures [10]. Accordingly, this book framed Modernity/Coloniality as the main source of injustice in the Anthropocene. This overlapping system of power produced core–periphery arrangements transforming othered peoples into invisible, inaudible, and illegible obstacles to progress. Consequently, this book posits a need to frame epistemic injustices as an essential component of humanity's multiple crises. The next section explores this concept and provides a vision of what epistemic justice entails in transition processes.

Epistemic justice

There is a clear intention in just transitions to engage with marginalised peoples by recognising and including plural identities. Scholars constantly point to issues of misrecognition, where certain groups are not treated as 'valid and respected members of society' [11, p. 443], an issue often intensified by legitimating the views of dominant social groups while ignoring others. Indeed, 'sustainability projects and policies often neglect lay knowledge, residents' land use preferences, communities' distinct cultural norms, common use of resources, sacred spaces, and diverse knowledge systems' [11, p. 444].

In line with these concerns, sustainability transitions (ST) research agendas call for working with non-experts and Indigenous communities as sources of knowledge [9], and 'examining which knowledges, practices, or testimonies are allowed to surface and are legitimated – without being captured or co-opted along the way – and which ones remain overlooked or pushed aside' [5, p. 1750].

It is important to recognise that this is not only a concern of ST, as engagements with justice are also brought up in broader socio-technical questions [13]. Additionally, issues of epistemic injustice have been explored extensively in critical theory, feminism, and decolonial and other scholarships [14]. In myriad other fields, the 'ability to be treated equally as a knower … [has] become increasingly recognised as an important aspect of justice' [15, p. 116].

Along with deepening notions of injustices in recognition and participation, the concept of epistemic injustice stems from a recognition that 'knowers can be harmed or wronged in their capacity as knowers' [14, p. 16]. These are distinct types of wrongs, where someone is downgraded/disadvantaged as an epistemic subject [16]. Epistemic injustice, then, 'is the idea that [people] can be unfairly discriminated against in [their] capacity as a knower based on prejudices about the speaker, such as gender, social background, ethnicity, race, sexuality, tone of voice, accent and so on' [15, p. 116]. Use of this concept should be strict, avoiding its expansion to interpersonal manipulation or 'systemic riggings of the epistemic economy', e.g., in unequal access to education [16, p. 53].

Engaging with these kinds of injustices can be highly problematic, as there is risk of falling for epistemic relativism. Scholars propose four lenses to understand its different varieties: (i) injustices that structure inequitable relations between knowers, and that maintain knowledge in the service of dominance; (ii) relational injustices that hinder knowledge relations and break trust; (iii) injustices maintained by knowledge systems working with efficacy; and (iv) injustices related to epistemic labour that hinder knowers, devalue certain types of knowers, and extract labour related to knowledge production [14].

Other scholars advocate for an engagement with epistemic injustice at the micro-level of interpersonal relations, defining two types of epistemic injustice: testimonial injustice, as in the shifting credibility of a statement based on prejudice against the speaker; and hermeneutical injustice, as a failure of shared intelligibility [15], [16].

Beyond the different typologies and scales of epistemic injustice, there might be specific conditions that overarch them. For Byskov [15], someone is unjustifiably discriminated against as a knower if the following conditions are met:

- The disadvantage condition: They must suffer epistemic and/or socio-economic disadvantages and inequalities resulting from the discrimination.
- The prejudice condition: The discrimination must involve prejudiced (i.e., unfair) sentiments about the speaker.
- The stakeholder condition: They must be somehow affected by the decisions that they are excluded from influencing.
- The epistemic condition: They must possess knowledge that is relevant for the decision that they are excluded from.
- The social justice condition: They must at the same time also suffer from other social injustices.

This book is particularly interested in epistemic injustices embedded in political institutions borne by the colonial power matrix. This resonates with framings of 'epistemic contracts' in service of an unjust social contract where system beneficiaries remain ignorant of unjust arrangements and end up developing epistemic arrogance or laziness [14]. This manifests in many aspects, from Euro-American philosophical traditions that set the bar by which all experiences of reality are measured, to 'racial contracts' creating ideal knowers and sub-knowers, or patriarchy framing women as 'less competent knowers' [14, p. 17]. Pohlhaus defines this as wilful hermeneutical ignorance, 'the propensity to dismiss whole aspects of the experienced world by refusing to become proficient in the epistemic resources required for attending to those parts of the world well' [14, p. 17].

This book is aligned with the posture that 'there [cannot be] global social justice without global cognitive justice' [17, Introduction]; as such, it wants to give special attention to epistemic justice. This is why it began with a recognition of Modernity/Coloniality as a structuring force and has explored interculturality and dialogue of knowledge(s) as a pathway for the recognition, validation, and inclusion of alternative historically marginalised ways of being.

Although there are no set definitions for the concept of cognitive justice, this work deviates from ideas of cognitive justice as 'more equal distributions of scientific knowledge' [17, Ch. 7]. It understands epistemic justice as a movement from concealing, ignoring, and eroding non-Modern knowledge(s), to a Transmodern solidarity-based epistemology where mutual education and critique between knowledge(s) is emphasised while prioritising those least advantaged [18]. While recognising that it is a slippery slope, it is important to emphasise that this is not a surrender to relativism, post-truth, or alternative facts, but a recognition of different lifeworlds and types of knowledge(s) [10], [18], [19].

Sustainability transitions occur under the overarching structure of Modernity/Coloniality, which restricts the possibilities open to, accessible by, or closed for people. Modernity/Coloniality has increased power and opportunities for those who align with its racist, patriarchal, heteronormative, ableist, and capitalist core, while restricting opportunities for all others. In this sense, Coloniality is unjust and produces multiple reinforcing injustices. Although no single actor can be blamed for structural injustices, there are some that are more powerful and thus have opportunities to challenge the structure.

Justice in Andean seed system transitions

The historical configuration of Andean seed systems has produced and perpetuated multiple injustices towards informal(ised) actors. The distribution of benefits and harms is skewed, informal system actors have only recently been recognised as valuable for the system, and they have been subjected to unfair processes by being placed in an a-legal space. Beyond the distributional, recognitive, and procedural justice triad, informal(ised) actors have been constantly excluded from decision-making spaces (participative injustice) and have been heavily impacted by industry (i.e., eroding their possibilities for subsistence) and climate change (environmental injustices). These impacts inevitably lead to intergenerational injustices, as young farmers see no possibilities of living a good life in rural contexts.

Informal(ised) system actors have also been historically harmed in their capacity as knowers; knowers of their territory, their seed rearing practices, and as producers of quality seeds produced by their knowledge(s). This epistemic harm is multidimensional and stems from entrenched prejudices about the knowers (e.g., being framed as obstacles to progress). These peoples also suffer from epistemic and socioeconomic disadvantages that result in multiple social injustices. One of these disadvantages is their exclusion from decision-making spaces determining how the system functions. This exclusion happens despite informal(ised) system actors holding knowledge relevant for the decisions that directly affect them. Additionally, it blocks possibilities for them to decide on benefit and burden distribution.

This book argues that the transitions led by seed guardian networks (SGNs) are fair and expand justice possibilities for historically marginalised actors. Transitions observed in Colombia, Ecuador, and Venezuela evidence the recognition of diverse value and knowledge systems. It can be said that it is an attempt to kickstart bold sustainability action through novel modes of governance. Although there is no evidence of just outcomes for informal(ised) system actors yet, the process has strived for their recognition, resulting in Transmodern legal frameworks that might enable their practices to thrive.

Increased legal protection for informal(ised) actors allows for better distribution of benefits across the socio-technical system. It is important to recognise that all justice processes produce winners and losers, so a Transmodern transition might lead to reduced opportunities for other actors. Although the legal frameworks produced in the Andean region maintain the formal and informal(ised) seed system, broader ambitions of food sovereignty might lead to weakening of the formal system. A key challenge to stabilise the emergent transition pathway will be to attend to the narratives, practices, and rules being phased out.

Andean seed system transitions can also be considered liberating processes, as they clearly see the abyssal line, open translational contact zones, and create Transmodern institutions. It could be argued that these processes not only liberate human actors but also their non-human counterparts. Native seeds are recognised and allowed to walk freely without owners, narratives of their agency are reproduced, and care practices towards them are institutionalised and protected by the state. As such, this is also a process that encourages ecological justice by changing power relations between humans and non-humans.

These transition processes have also been highly intersectional. They have led to the acknowledgement and inclusion of plural identities, especially those that have been made invisible by Coloniality. SGN dynamics also opened spaces of permanent intercultural dialogue and connected plural social movements that include gender-based, class-based, and environmental agendas. Ultimately, they produced intersectional institutions emerging from legislative changes. Accordingly, these transition processes have also been highly relational, starting from the politicisation of the everyday by seed guardians, the networked nature of the niche, and developing sustainable interventions that opened spaces for other ways of being.

Andean seed system transitions have been Transmodern sustainability transitions. These processes require epistemic justice to open pathways that move whole systems (and hopefully all of society) from concealing, ignoring, and eroding non-Modern knowledge(s), to a solidarity-based epistemology where mutual education and critique between knowledge(s) is emphasised while prioritising those least advantaged.

In the cases explored, marginalised actors and their diverse knowledge(s) are legitimated, allowing them to be treated equally as knowers. Although informal(ised) system actors are still wary of co-optation by the dominant food regime, their narratives and practices are now more protected by the state. This might put a stop to their constant downgrading as epistemic subjects, reducing instances of discrimination.

Transmodern seed system transitions might lead to more equitable relations between knowers, and the mobilisation of othered knowledge(s) towards challenging Modernity/Coloniality. These processes might also move the system from relations of distrust between formal and informal(ised) actors towards more solidarity and innovation across the abyssal line. The take-off phase of this transition (novel Transmodern legislation and micro-institutions) might also recognise, map, and value the wide variety of epistemic labour in the system, leading to new practices of knowledge production across the abyssal line.

This chapter has argued that Andean seed system transitions contribute to resolving structural injustices and promote multidimensional justice. Crucially, one of these dimensions is food justice, or 'the adequate supply of and access to decent food but also community autonomy and self-determination and thus community control over their own food resources and practices' [2, p. 13]. Although power is now more distributed in the system, this will not guarantee the achievement of food security at national level. All system actors should be attentive to the ultimate purpose of keeping people well fed, which necessarily implies hybrid system configurations.

Conclusion

In this chapter, multiple intersecting justice types were used to analyse Andean seed system transitions. These processes are unfolding within the colonial matrix of power, an inherently unjust structure that organised the world in core–periphery dynamics. This structure manifests in seed systems as an abyssal line that divides them by validating some practices and knowledge(s) while ignoring others. This produces multiple injustices: exclusion of informal(ised) actors from decision-making, erosion of their possibilities for autonomous subsistence, and forced outmigration, among others.

Injustices related to knowledge were given special attention. Despite their importance for food security in the region, the intellectual labour of informal(ised) actors appears to be invisible to incumbents. This is a product of colonial epistemic hierarchies that have marginalised AfroIndopeasant ways of being and understanding the world for over 500 years. These power asymmetries have led to their exclusion from decision-making, despite holding relevant knowledge and being directly affected by decisions made by incumbents.

Over the last 30 years, relations between informal(ised) actors and the state have transformed. Macro-phenomena like the climate emergency and the crisis of the dominant food regime have made alternatives more appealing. Responding to these crises, diverse actors have configured large-scale networks, enacting alternatives informed by *Buen Vivir* and food sovereignty. These simultaneous pressures have led to new spaces of encounter enabled by states more open to direct citizen participation.

The co-creation of inclusive Transmodern seed policies and institutions represents a significant reconfiguration of knowledge, power, and governance. This process expanded participation, access to information, and legal protections. It redistributed authority in knowledge production, as states now recognise AfroIndopeasants and seed guardians as knowledge-holders. Andean states have enriched the parameters by which they decide what constitutes valid knowledge and what is a knower in the system. As such, epistemic justice is emerging by moving from concealing, ignoring, or eroding Transmodern knowledge towards the establishment of mosaic ways of knowing.

These developments suggest the emergence of a just transition pathway in Andean seed systems. New configurations enabled by Transmodern seed policy not only expand participation and innovation, but may impact food, environmental, ecological, and intergenerational justice. Although just outcomes are not guaranteed and these transitions remain fragile and contested, Modern/Colonial relational patterns have been substantially altered.

Sustaining these Transmodern transitions will be challenging; it will require constant attention to power, knowledge, and innovation while shielding them from co-optation. However, this book argues that transition pathways toward more just, plural, and sustainable food futures have emerged in Andean seed systems. Beyond Latin America, similar processes are underway (see [20] for a case in India). Decolonisation might take 500 years and thinking in such timeframes is discouraging. However, Modernity/Coloniality is increasingly showing its cracks, and peoples around the world are innovating in Transmodern terms to bring forth more just and sustainable futures.

Note

1 However, this book aligns with the position that past imperial powers, rich nations, and particularly the richest 10 per cent of people in those nations, hold more responsibility than others in current social and environmental structural injustices.

References

[1] D. A. Heyen, 'Social justice in the context of climate policy: Systematizing the variety of inequality dimensions, social impacts, and justice principles', *Climate Policy*, vol. 23, no. 5, pp. 539–554, 2023, doi:10.1080/14693062.2022.2142499.

[2] E. Gilson and S. Kenehan, 'Introduction', in *Food, environment, and climate change justice at the intersections*, E. Gilson and S. Kenehan, Eds. Lanham, MD: Rowman & Littlefield International, 2021, pp. 1–26.

[3] J. McGregor, 'The intersection of environmental, climate, and food justice', in *Food, environment, and climate change justice at the intersections*, E. Gilson and S. Kenehan, Eds. Lanham, MD: Rowman & Littlefield International, 2021, pp. 29–46.

[4] T. Kortetmäki, 'Nobody's fault? structural injustice, food, and climate change', in *Food, environment, and climate change justice at the intersections*, E. Gilson and S. Kenehan, Eds. Lanham, MD: Rowman & Littlefield International, 2021, pp. 47–62.

[5] I. Anguelovski *et al.*, 'Expanding the boundaries of justice in urban greening scholarship: Toward an emancipatory, antisubordination, intersectional, and relational approach', *Ann Am Assoc Geogr*, vol. 110, no. 6, pp. 1743–1769, 2019.

[6] E. Morena, D. Krause, and D. Stevis, 'Introduction: The genealogy and contemporary politics of just transitions', in *Just transitions: Social justice in the shift*

towards a low-carbon world, E. Morena, D. Krause, and D. Stevis, Eds. London: Pluto Press, 2019.

[7] M. C. Rawlinson, 'The climate of food: Justice, truth, and structural change', in *Food, environment, and climate change justice at the intersections*, E. Gilson and S. Kenehan, Eds. Lanham, MD: Rowman & Littlefield International, 2021, pp. 91–118.

[8] C. Loo, 'Participation and food justice in light of global climate change', in *Food, environment, and climate change justice at the intersections*, E. Gilson and S. Kenehan, Eds. Lanham, MD: Rowman & Littlefield International, 2021, pp. 63–76.

[9] S. Williams and A. Doyon, 'Justice in energy transitions', *Environ Innov Soc Transit*, vol. 31, pp. 144–153, 2019, doi:10.1016/j.eist.2018.12.001.

[10] S. Ryder, K. Powlen, M. Laituri, S. A. Malin, J. Sbicca, and D. Stevis, *Environmental justice in the Anthropocene: From (un)just presents to just futures.* London: Routledge, 2021.

[11] D. Ciplet and J. L. Harrison, 'Transition tensions: Mapping conflicts in movements for a just and sustainable transition', *Env Polit*, vol. 29, no. 3, pp. 435–456, 2020, doi:10.1080/09644016.2019.1595883.

[12] M. Swilling, J. Musango, and J. Wakeford, 'Developmental states and sustainability transitions: Prospects of a just transition in South Africa', *Journal of Environmental Policy and Planning*, vol. 18, no. 5, pp. 650–672, Oct.2016, doi:10.1080/1523908X.2015.1107716.

[13] K. Moore, N. Hernández Vidal, and D. L. Kleinman, 'Knowledge and justice: A comment', *Science and Culture*, vol. 28, no. 3, pp. 383–390, 2019. doi:10.1080/09505431.2019.1649646.

[14] G.PohlhausJr., 'Varieties of epistemic injustice', in *The Routledge handbook of epistemic injustice*, 1st ed., I. James Kidd, J. Medina, and G.PohlhausJr., Eds. London: Routledge, 2017, pp. 13–26. doi:10.1007/s11229-020-02634-y.

[15] M. F. Byskov, 'What makes epistemic injustice an "injustice"?' *J Soc Philos*, vol. 52, no. 1, pp. 114–131, 2021, doi:10.1111/josp.12348.

[16] M. Fricker, 'Evolving concepts of epistemic injustice', in *The Routledge handbook of epistemic injustice*, 1st ed., I. James Kidd, J. Medina, and G.PohlhausJr., Eds. London: Routledge, 2017, pp. 53–60. doi:10.4324/9781315212043.

[17] B. de S. Santos, *Epistemologies of the South – Justice against epistemicide.* New York: Routledge, 2016.

[18] B. Reiter, Ed., *Constructing the pluriverse.* Durham, NC: Duke University Press, 2018.

[19] A. Escobar, *Designs for the pluriverse: Radical interdependence, autonomy, and the making of worlds.* Durham, NC: Duke University Press, 2018.

[20] P. Singh, 'Politics of knowledge in development: Explorations in seed sovereignty', *Studies in Indian Politics*, vol. 9, no. 1, pp. 105–117, Jun.2021, doi:10.1177/2321023021999179.

Conclusion

Calls to attend to the structural factors that maintain our food systems in crisis have grown louder in recent decades. Many researchers and practitioners have mapped these factors and contributed visions for food futures. However, knowledge on how to operationalise transformation remains comparatively scarce, particularly as it pertains to more radical alternatives that aim to tackle power asymmetries and achieve more sustainable configurations.

This book seeks to contribute to this gap by synthesising lessons from Andean seed system transformations in the early 21st century. Its focus is on multi-actor processes that have expanded system governance to include historically marginalised actors. The intention of this book is not to propose a universal template to produce change in food systems but rather to define the contours of actual transformative processes. The hope is that by mapping these dynamics, others may adapt these insights to their own contexts to better govern food system transitions.

This book interpreted current crises (environmental, social) of the food system not as isolated problems but as a result of Modernity/Coloniality. As Modernity's exhausted promises become increasingly evident, transformations to more regenerative and pluriversal alternatives are needed. Accordingly, the book directed attention to counterhegemonic narratives of wellbeing and innovative practices enacted on the margins of Western(ised) society.

As ecological and social crises intensify, dominant Western frames destabilise, creating a fragile space for Transmodern innovations to flourish. Fully decolonising our societies might take hundreds of years. However, the processes described in this book show ways in which people chip away from Modernity/Coloniality to expand its cracks. This occurs through everyday actions of sustainable materialism, institutional experiments, and collective political engagement, effectively operationalising decolonisation.

The book aimed to explore emerging Transmodern innovative spaces that enabled creative tension between non-Modern and Modern/Colonial worldviews. Seed guardian networks (SGNs) provided a fertile ground for exploring these dynamics. As decentralised, diverse, and trans-local coalitions, SGNs leveraged alternative narratives, strengthened border practices, and challenged Modern assumptions of seed knowledge and governance. Their actions reveal

DOI: 10.4324/9781003715092-14

how alternative worldviews can be enacted and how plural traditional knowledge(s) can become legitimate drivers of system change.

The first objective of this book was to engage critically with Modernity to uncover the persistence of Coloniality. Part 1 explored theoretical developments by scholars working in the Latin American decolonial option, Transmodernity, and *Buen Vivir*. This conceptual framing was used to analyse the current state of global and Latin American seed systems, uncovering how they are marked by the abyssal line and abyssal thinking. The book established Modernity/Coloniality as that which needs to be transformed, since Western ways of being and knowing are in crisis. Conversely, Transmodernity and *Buen Vivir* were positioned as desired futures capable of addressing current challenges through relationality and translational contact zones.

Contemporary seed systems are divided into a formal/industrial and informal(ised)/traditional system. This division stems from distinct worldviews that inform radically different knowledge, practices, materials/technology, and rules. The Modern formal/industrial system holds great power as it is seen as the only true path towards global food security. However, in poor postcolonial nations, the informal(ised)/traditional system still produces much of the food humans consume. It is nonetheless constantly ignored and devalued as it is still judged from entrenched Modern knowledge hierarchies. Seen through a Transmodern lens, the informal(ised) system has the potential to produce more sustainable and just food systems informed by postures of interrelatedness, cooperation, participation, and empowerment.

The second objective of this book was to describe the internal dynamics of seed guardian networks and their interactions with other system actors (see Part 2). This led to an understanding of SGNs as horizontal organisations with distributed leaderships, a highly diverse membership, and transnational linkages. The networks were positioned as spaces of dialogue between knowledge(s) and experimentation with agri-food alternatives. SGNs are a particular type of social organisation, perhaps without clear agendas and programmes, but transformative because of its material everyday practices. Their persistent place-based trans-local experimentation was transformative, as it gradually expanded the boundaries of what is considered legitimate or possible.

Regarding interactions with incumbent actors, the book focused on: (i) state informalisation of traditional seed practices as part of Coloniality (i.e., making them invisible); and (ii) how these networks were able to transcend invisibilisation to become influential players in the creation of new seed system rules. Additionally, interactions between seed guardians and other informal(ised) food system actors revealed profound differences between them. These tensions reinforced the idea that SGNs are not just traditional grassroots actors but Transmodern innovators who blur the distinctions between the formal and informal, the scientific and the ancestral, the Modern and the non-Modern.

The third objective was to document the trajectories of national seed system transformations in Colombia, Ecuador, and Venezuela. Chapters 4–6 provided thick descriptions of political dynamics, legislative processes, and institutional reforms over the last three decades. In all cases, SGN activities were central to transformations as their practices became more visible over time.

In the case of Colombia, a marked divide across the abyssal line preserved tensions between the formal and informal(ised) systems exacerbated by the signing of international trade partnership agreements. These tensions came to a head with an abnormal rainy season coinciding with the implementation of an inconsistent seed law in 2010. After the signing of a peace accord with the FARC-EP guerrilla organisation in 2016, both sides of the abyssal line began to come closer as timid *Buen Vivir* framings led to more spaces for family agriculture, food sovereignty, and agroecology. This promoted cultural shift in agricultural authorities, leading to a participatory process to define the contours of a new national seed system.

In the case of Venezuela, the Bolivarian Revolution was framed as a transition from capitalist extractivism to a participatory socialism where humans and nature coexist harmoniously. The resulting Bolivarian constitution of 1999 was framed as eco-socialist and directed the national food system closer to agroecology and food sovereignty. This opened spaces to cross the abyssal line, leading to processes of direct democracy where people could contribute directly to establishing system rules. As a new seed law started being discussed in 2013, a space of popular debate was opened by informal(ised) actors to include their agendas. In 2015, four drafts of the law were presented to the National Assembly, each including the perspectives of different system actors. Due to pressures from SGNs and other social movements, a law was approved recognising a formal/informal hybrid seed system. This law ordered the creation of a Popular Council for the Safeguarding and Protection of Local, Peasant, Indigenous and Afro-descendant Seed and kickstarted a Scientific-Peasant Alliance, which encouraged horizontal dialogues between plural knowledge(s). Crucially, it recognised local knowledge-holders as *maestres pueblo*, exalting the value of their traditional knowledge.

In the case of Ecuador, the emergence of the Quichua neologism *Sumak Kawsay* in the early 1990s opened a space to cross the abyssal line. This narrative of *Buen Vivir*/good living expanded from the Amazon to every corner of Latin America, informing contemporary politics in Ecuador and beyond. After decades of political turmoil, a 21st century socialist project emerged that was aimed at tackling the crisis. Once in power, this *movimientista* left set out to implement a citizen's revolution that overhauled participation and directed political institutions towards post-Development. This produced a *Buen Vivir* constitution (2008) that recognised the rights of nature and rights of people to food sovereignty. The post-constitutional Organic Law of Food Sovereignty opened a participatory process inclusive of historically marginalised voices. As a result, the Organic Law of Agrobiodiversity, Seeds, and Promotion of Agriculture was approved in 2017, recognising the role of native seeds and associated ancestral knowledge, sustainable agriculture, and peasant families in the Ecuadorian seed system.

The final objective of this book was to contrast processes of Andean seed system transformation with existing theories of change (see Part 3). Transformation occurred as translational contact zones were opened to negotiate boundaries between worldviews. This enabled the emergence of novel seed system rules incentivising institutional entrepreneurs (within SGNs and the state) to design new micro-institutions through processes of bricolage.

Chapter 7 framed change as occurring across narratives, practices, and rules. Theories of narrative transformation were mobilised to better understand the formation of discourse coalitions. Food sovereignty, *Buen Vivir*, and a new plural left provided meaning to the process. These narratives envision a post-capitalist or at least post-neoliberal post-Development and a wider move towards biocentricity. Participation is paramount, as direct democracy is favoured, inclusion is demanded, and plurality is encouraged. Ideas of interconnectedness, diversity, and partnership feature heavily, as well as an expansion of rights inclusive of nature and a push for sustainability. These shared ideational elements linked diverse actors in SGNs, civil society, and the state.

Regarding practices, the scaling up of seed guardianship and the opening of citizen participation in policymaking were critical. These simultaneous processes led to the creation of different system rules, a shift that enabled novel practice configurations that are more Transmodern. For instance, state technicians are now developing competences to care for native seeds or support the creation of seed houses and exchanges.

Chapter 8 analysed the cases using sustainability transitions theory and a decolonial interpretation of the multi-level perspective. A multi-level reading of Andean seed system transformations evidenced macro-level trends and emerging countertrends, informing change at regime and niche levels. The regime was described as a Transmodern hybrid between Modern and non-Modern worldviews divided by the abyssal line. SGNs were positioned as Transmodern grassroots innovators capable of destabilising hybrid regimes. These niche spaces were initially shielded, nurtured, and empowered from within. As they have been increasingly included in legal frameworks, states appear to be more involved in strengthening them.

The SGN niche is now seen and is being nurtured by the state due to the emergence of translational contact zones. This emergence was explained as the articulation of diverse intermediary roles, creating momentum for change by linking niche actors, resources, visions, and demands with the existing regime. To conclude the analysis from a sustainability transitions perspective, Chapter 9 argued that Andean seed system change can be interpreted as a just transition. It allowed epistemic justice to emerge, distributing benefits more evenly in the socio-technical system, there is now more access to information, distributed governance, and legal protection for alternative practices.

How did seed system transformations occur in Colombia, Ecuador, and Venezuela in the early 21st century across the abyssal line? This book has argued that it was through the opening of translational contact zones by actors invested in transcending the abyssal line. These spaces for dialogue between diverse knowledge(s) emerged from different pressures and opportunities in seed systems, including:

- Macro-level trends related to the crisis of Modernity, consolidation of neoliberalism, and the multidimensional crisis of the dominant agri-food system. On the other hand, there were emerging narratives like Transmodernity, the pursuit of post-capitalist and post-Development alternatives, as well as an agroecological transition towards food sovereignty. Narrative spread was supported by the global diffusion of mobile devices, digitalisation, and social media.
- Surviving non-Modern border practices being revalued and reconfigured by innovative actors engaging in dialogues with Modernity (i.e., creating Transmodern practices); and diffusing these practices through trans-local networks that attract new practitioners and making visible that which was made invisible by Modernity/Coloniality.
- Shifting state practices towards more direct democracy, allowing historically marginalised actors to be part of decision-making processes.
- The articulation of discourse coalitions through shared ideational elements related to *Buen Vivi*r, food sovereignty, and a new plural left. These articulations allowed intermediaries in SGNs and the state to connect, make sense of the system, and facilitate systems change.
- A pursuit of epistemic justice, especially by SGNs and other informal(ised) system actors (re)claiming their knowledge(s) as valuable for national seed systems. On the other hand, there were incumbents recognising the importance of the informal(ised) system in national food provision.

The emergence of translational contact zones led to recognition of reciprocal incompleteness, allowing the expansion of parameters that define authorship, intelligibility, and purposefulness. Accordingly, system rules have shifted to allow more space for non-Modern and Transmodern knowledge(s) and practices. SGNs went from being rule-takers kept in a non-legal space to rule-makers recognised for their contributions to the system. Their knowledge is now exalted by governments and is often framed as equivalent to Modern technoscience. Additionally, new micro-institutions have emerged, allowing for more inter-epistemic dialogues between worldviews. As such, the legal and scientific abyssal lines have been crossed, furthering epistemic justice in the system. This may lead to a solidarity-based epistemology where multiple knowledge(s) engage in dialogue while prioritising those least advantaged.

It is important to recognise that this book has a positive results bias. Negative aspects of this transition have not been sufficiently acknowledged. The persistence of colonial epistemic asymmetries still frames informal(ised) actors as obstacles to progress despite their relevance for food production in Latin America. Native seeds are still perceived as being poor-quality raw materials, ignoring the knowledge and labour that goes into their conservation. Additionally, participation and inclusion of marginalised actors in seed system rules are seen as counterproductive by some incumbents. Transnational corporations still wield enormous power over seed systems and global regulations are still skewed towards industrialism.

The unrelenting force of Modernity puts this emergent Transmodern transition pathway in a precarious position. However, marginalised actors keep challenging the Modern by materialising alternative futures to Western development. Mainstreaming of these alternatives at global scale might be far away, but their linkage might destabilise the Modern/Colonial Anthropocene. This book has argued that Transmodern transitions to *Buen Vivir* have emerged in early 21st century Andean seed systems. The implementation of new system rules inclusive of SGN concerns established a more level playing field between formal and informal(ised) seed systems, signalling the beginning of a take-off phase towards more pluriversal biocentric futures.

Scholars have drawn attention to the revisionist nature of decolonial approaches to transitions, as they usually start from Western modes of thinking. Regrettably, this work may be criticised as such, as it could have used a more explicit inductive approach. However, Modern ways of knowing should not be irreflexively discarded in the name of emancipation, liberation, or decolonisation. This is an essential part of Transmodern dialogue. Transmodernity opens an interesting space for reflection as it intends to move beyond First-World and Third-World fundamentalisms. This posture should inform a future research agenda on Transmodern transitions to pluriversal biocentric futures that could prioritise the following items:

- Mapping other Transmodern innovative initiatives worldwide with the aim of developing comparative analyses between them.
- Understanding the governance of those innovative spaces, including their upscaling dynamics.
- Observing the emergence of translational contact zones as Transmodern innovators interact with incumbents.
- Evaluating if Transmodern legislative change effectively leads to the stabilisation of Transmodern transitions.
- Exploring consensus and conflict dynamics between multiple system actors as they engage in knowledge interactions across the abyssal line.
- Observing these processes through other theoretical lenses and attempting to theorise about them in bottom-up ways.

This research agenda is also an invitation to other researchers, practitioners, and policymakers to further examine food systems with process theories of change through a decolonial lens. Discourse theory, institutional theory, social practice theory, and sustainability transitions have shed light on transformation processes in this book. However, there is still much to uncover in these and other theories, particularly for poor postcolonial nations that face persistent barriers in knowledge production. Post-Development and post-capitalist strands of transformations research are formidable when critiquing current systems or proposing visions. However, they need deeper engagements with how change happens in practice. The case studies presented here are Transmodern enactments that elucidate ways to operationalise transitions towards biocentric societies.

The message for states and development agencies is to push for the establishment of hybrid food systems. The formal/industrial and informal(ised)/traditional system can coexist, as they have done for more than a hundred years. However, faced with the daunting crises of the Anthropocene, humanity needs to phase out degenerative systems and stimulate regenerative ones. This entails the creation of conditions for alternative practices to thrive without subsuming them into Modern, capitalist, or industrialist visions. This requires assuming a Transmodern worldview that aims to integrate the best of the pre-Modern, and (post)Modern. It means treating alternative practices as socio-technical innovations, micro-scale experiments in living otherwise that need nurturing and shielding.

This book promoted non-Modern visions of sustainable futures created by historically marginalised actors. It advocated for their inclusion in established theories while being mindful of power and justice. The intention here is to advance plurality of knowledge(s) in a time when there seem to be no Modern solutions to Modern problems. As such, this book is framed as a device of epistemic direct action, challenging theories from the experiences of those historically marginalised by Coloniality. Hopefully, it has shown that Transmodern transitions to *Buen(os) Vivir(es)* are desirable and possible. Crucially, it calls for more attention to the micro-level, the potency of micro-utopias, and the power of oppositional micro-institutions to alleviate planetary crises.

Andean seed guardian networks and state intermediaries pushing for alternatives have evidenced how to transform the food system. The pathway they have set up could move it towards being more sustainable and just. The seeds of change they have planted need to be cared for by all to transform the dominant agri-food regime. As guardians speak of native seeds, may their knowledge circulate freely and reach all corners of the world.

Index

Note: Page numbers in *italics* indicate figures, **bold** indicate tables in the text, and references following "n" refer to the notes.

For Product Safety Concerns and Information please contact our EU
representative GPSR@taylorandfrancis.com
Taylor & Francis Verlag GmbH, Kaufingerstraße 24, 80331 München, Germany

www.ingramcontent.com/pod-product-compliance
Ingram Content Group UK Ltd.
Pitfield, Milton Keynes, MK11 3LW, UK
UKHW020142200926
475702UK00005B/876

* 9 7 8 1 0 4 1 2 0 1 1 1 3 *